ELEMENTARY STATISTICAL TABLES

F.D.J. DUNSTAN, M.A., M.Sc., D.Phil.
A.B.J. NIX, B.Sc., Ph.D.
J.F. REYNOLDS, M.A., Ph.D.
R.J. ROWLANDS, B.Sc., Ph.D.

RND PUBLICATIONS

PREFACE

This set of tables has been designed by RND Publications in collaboration with the Associated Examining Board for use in Advanced level and university courses in Statistics. Each table is preceded by a brief explanation of its contents and we hope that, in general, the lay-out is sufficiently familiar to enable the tables to be used satisfactorily without further explanation. We have resisted the temptation to include excessive material on the use of tables and we leave this to the textbooks. Furthermore, since the tables will be used in examinations, many of the formulae which are expected to be known by the candidates have also been omitted.

We have tried to maintain a degree of consistency in presentation. To this end, we have tabulated the distribution function in Tables 1, 2 and 3. In Tables 4, 6, 7 and 8, the percentage points are tabulated since these distributions are used in many different ways. In Tables 9, 10, 13 and 14, the upper tail critical values are tabulated since the corresponding distributions are used almost exclusively in hypothesis testing.

In Tables 1 and 2, for ease of presentation, as soon as the value of the distribution function reaches unity, all succeeding ones are omitted. Thus if, in using these tables, a blank is obtained as the required probability, this should be interpreted as unity.

In Tables 10, 13 and 14 where non-parametric discrete statistics are tabulated, the values given should be included within the critical region. Furthermore, as explained in the headings, exact significance levels cannot in general be obtained using these statistics. The critical values given are those which ensure a significance level as close as possible to the stated levels. If, in using these tables, a blank is obtained as the required critical value, this means that the nearest achievable significance level to the stated level is 0%. The corresponding critical value is omitted on the grounds that such a test has no practical value.

Published by RND PUBLICATIONS
The Maltings, East Tyndall Street,
Cardiff CF1 5EA.

ISBN 0 - 9506719 - 5 - 9

Second Edition 1988

CONTENTS

TABLE 1 BINOMIAL DISTRIBUTION FUNCTION

The table gives the probability of obtaining at most x successes in a sequence of n independent trials, each of which has a probability p of success, i.e.

$$P(X \leq x) = \sum_{r=0}^{x} \binom{n}{r} p^r (1-p)^{n-r}$$

where X denotes the number of successes.

n	x \ p	0.01	0.02	0.03	0.04	0.05	0.06	0.07	0.08	0.09	0.10	0.15	0.20	0.25	0.30	0.35	0.40	0.45	0.50	p / x
n = 2	0	.9801	.9604	.9409	.9216	.9025	.8836	.8649	.8464	.8281	.8100	.7225	.6400	.5625	.4900	.4225	.3600	.3025	.2500	0
	1	.9999	.9996	.9991	.9984	.9975	.9964	.9951	.9936	.9919	.9900	.9775	.9600	.9375	.9100	.8775	.8400	.7975	.7500	1
	2	1.000	1.000	1.000	1.000	1.000	1.000	1.000	1.000	1.000	1.000	1.000	1.000	1.000	1.000	1.000	1.000	1.000	1.000	2
n = 3	0	.9703	.9412	.9127	.8847	.8574	.8306	.8044	.7787	.7536	.7290	.6141	.5120	.4219	.3430	.2746	.2160	.1664	.1250	0
	1	.9997	.9988	.9974	.9953	.9928	.9896	.9860	.9818	.9772	.9720	.9393	.8960	.8438	.7840	.7183	.6480	.5748	.5000	1
	2	1.000	1.000	1.000	.9999	.9999	.9998	.9997	.9995	.9993	.9990	.9966	.9920	.9844	.9730	.9571	.9360	.9089	.8750	2
	3				1.000	1.000	1.000	1.000	1.000	1.000	1.000	1.000	1.000	1.000	1.000	1.000	1.000	1.000	1.000	3
n = 4	0	.9606	.9224	.8853	.8493	.8145	.7807	.7481	.7164	.6857	.6561	.5220	.4096	.3164	.2401	.1785	.1296	.0915	.0625	0
	1	.9994	.9977	.9948	.9909	.9860	.9801	.9733	.9656	.9570	.9477	.8905	.8192	.7383	.6517	.5630	.4752	.3910	.3125	1
	2	1.000	1.000	.9999	.9998	.9995	.9992	.9987	.9981	.9973	.9963	.9880	.9728	.9492	.9163	.8735	.8208	.7585	.6875	2
	3			1.000	1.000	1.000	1.000	1.000	1.000	.9999	.9999	.9995	.9984	.9961	.9919	.9850	.9744	.9590	.9375	3
	4									1.000	1.000	1.000	1.000	1.000	1.000	1.000	1.000	1.000	1.000	4
n = 5	0	.9510	.9039	.8587	.8154	.7738	.7339	.6957	.6591	.6240	.5905	.4437	.3277	.2373	.1681	.1160	.0778	.0503	.0313	0
	1	.9990	.9962	.9915	.9852	.9774	.9681	.9575	.9456	.9326	.9185	.8352	.7373	.6328	.5282	.4284	.3370	.2562	.1875	1
	2	1.000	.9999	.9997	.9994	.9988	.9980	.9969	.9955	.9937	.9914	.9734	.9421	.8965	.8369	.7648	.6826	.5931	.5000	2
	3		1.000	1.000	1.000	1.000	.9999	.9999	.9998	.9997	.9995	.9978	.9933	.9844	.9692	.9460	.9130	.8688	.8125	3
	4						1.000	1.000	1.000	1.000	1.000	.9999	.9997	.9990	.9976	.9947	.9898	.9815	.9688	4
	5											1.000	1.000	1.000	1.000	1.000	1.000	1.000	1.000	5
n = 6	0	.9415	.8858	.8330	.7828	.7351	.6899	.6470	.6064	.5679	.5314	.3771	.2621	.1780	.1176	.0754	.0467	.0277	.0156	0
	1	.9985	.9943	.9875	.9784	.9672	.9541	.9392	.9227	.9048	.8857	.7765	.6554	.5339	.4202	.3191	.2333	.1636	.1094	1
	2	1.000	.9998	.9995	.9988	.9978	.9962	.9942	.9915	.9882	.9842	.9527	.9011	.8306	.7443	.6471	.5443	.4415	.3438	2
	3		1.000	1.000	1.000	.9999	.9998	.9997	.9995	.9992	.9987	.9941	.9830	.9624	.9295	.8826	.8208	.7447	.6563	3
	4					1.000	1.000	1.000	1.000	1.000	.9999	.9996	.9984	.9954	.9891	.9777	.9590	.9308	.8906	4
	5										1.000	1.000	.9999	.9998	.9993	.9982	.9959	.9917	.9844	5
	6												1.000	1.000	1.000	1.000	1.000	1.000	1.000	6
n = 7	0	.9321	.8681	.8080	.7514	.6983	.6485	.6017	.5578	.5168	.4783	.3206	.2097	.1335	.0824	.0490	.0280	.0152	.0078	0
	1	.9980	.9921	.9829	.9706	.9556	.9382	.9187	.8974	.8745	.8503	.7166	.5767	.4449	.3294	.2338	.1586	.1024	.0625	1
	2	1.000	.9997	.9991	.9980	.9962	.9937	.9903	.9860	.9807	.9743	.9262	.8520	.7564	.6471	.5323	.4199	.3164	.2266	2
	3		1.000	1.000	.9999	.9998	.9996	.9993	.9988	.9982	.9973	.9879	.9667	.9294	.8740	.8002	.7102	.6083	.5000	3
	4				1.000	1.000	1.000	1.000	.9999	.9999	.9998	.9988	.9953	.9871	.9712	.9444	.9037	.8471	.7734	4
	5								1.000	1.000	1.000	.9999	.9996	.9987	.9962	.9910	.9812	.9643	.9375	5
	6											1.000	1.000	.9999	.9998	.9994	.9984	.9963	.9922	6
	7													1.000	1.000	1.000	1.000	1.000	1.000	7
n = 8	0	.9227	.8508	.7837	.7214	.6634	.6096	.5596	.5132	.4703	.4305	.2725	.1678	.1001	.0576	.0319	.0168	.0084	.0039	0
	1	.9973	.9897	.9777	.9619	.9428	.9208	.8965	.8702	.8423	.8131	.6572	.5033	.3671	.2553	.1691	.1064	.0632	.0352	1
	2	.9999	.9996	.9987	.9969	.9942	.9904	.9853	.9789	.9711	.9619	.8948	.7969	.6785	.5518	.4278	.3154	.2201	.1445	2
	3	1.000	1.000	.9999	.9998	.9996	.9993	.9987	.9978	.9966	.9950	.9786	.9437	.8862	.8059	.7064	.5941	.4770	.3633	3
	4			1.000	1.000	1.000	1.000	.9999	.9999	.9997	.9996	.9971	.9896	.9727	.9420	.8939	.8263	.7396	.6367	4
	5							1.000	1.000	1.000	1.000	.9998	.9988	.9958	.9887	.9747	.9502	.9115	.8555	5
	6											1.000	.9999	.9996	.9987	.9964	.9915	.9819	.9648	6
	7												1.000	1.000	.9999	.9998	.9993	.9983	.9961	7
	8														1.000	1.000	1.000	1.000	1.000	8

x \ p	0.01	0.02	0.03	0.04	0.05	0.06	0.07	0.08	0.09	0.10	0.15	0.20	0.25	0.30	0.35	0.40	0.45	0.50	p / x
n = 9 0	.9135	.8337	.7602	.6925	.6302	.5730	.5204	.4722	.4279	.3874	.2316	.1342	.0751	.0404	.0207	.0101	.0046	.0020	0
1	.9966	.9869	.9718	.9522	.9288	.9022	.8729	.8417	.8088	.7748	.5995	.4362	.3003	.1960	.1211	.0705	.0385	.0195	1
2	.9999	.9994	.9980	.9955	.9916	.9862	.9791	.9702	.9595	.9470	.8591	.7382	.6007	.4628	.3373	.2318	.1495	.0898	2
3	1.000	1.000	.9999	.9997	.9994	.9987	.9977	.9963	.9943	.9917	.9661	.9144	.8343	.7297	.6089	.4826	.3614	.2539	3
4			1.000	1.000	1.000	.9999	.9998	.9997	.9995	.9991	.9944	.9804	.9511	.9012	.8283	.7334	.6214	.5000	4
5						1.000	1.000	1.000	1.000	.9999	.9994	.9969	.9900	.9747	.9464	.9006	.8342	.7461	5
6										1.000	1.000	.9997	.9987	.9957	.9888	.9750	.9502	.9102	6
7												1.000	.9999	.9996	.9986	.9962	.9909	.9805	7
8													1.000	1.000	.9999	.9997	.9992	.9980	8
9															1.000	1.000	1.000	1.000	9
n = 10 0	.9044	.8171	.7374	.6648	.5987	.5386	.4840	.4344	.3894	.3487	.1969	.1074	.0563	.0282	.0135	.0060	.0025	.0010	0
1	.9957	.9838	.9655	.9418	.9139	.8824	.8483	.8121	.7746	.7361	.5443	.3758	.2440	.1493	.0860	.0464	.0233	.0107	1
2	.9999	.9991	.9972	.9938	.9885	.9812	.9717	.9599	.9460	.9298	.8202	.6778	.5256	.3828	.2616	.1673	.0996	.0547	2
3	1.000	1.000	.9999	.9996	.9990	.9980	.9964	.9942	.9912	.9872	.9500	.8791	.7759	.6496	.5138	.3823	.2660	.1719	3
4			1.000	1.000	.9999	.9998	.9997	.9994	.9990	.9984	.9901	.9672	.9219	.8497	.7515	.6331	.5044	.3770	4
5					1.000	1.000	1.000	1.000	.9999	.9999	.9986	.9936	.9803	.9527	.9051	.8338	.7384	.6230	5
6									1.000	1.000	.9999	.9991	.9965	.9894	.9740	.9452	.8980	.8281	6
7											1.000	.9999	.9996	.9984	.9952	.9877	.9726	.9453	7
8												1.000	1.000	.9999	.9995	.9983	.9955	.9893	8
9														1.000	1.000	.9999	.9997	.9990	9
10																1.000	1.000	1.000	10
n = 11 0	.8953	.8007	.7153	.6382	.5688	.5063	.4501	.3996	.3544	.3138	.1673	.0859	.0422	.0198	.0088	.0036	.0014	.0005	0
1	.9948	.9805	.9587	.9308	.8981	.8618	.8228	.7819	.7399	.6974	.4922	.3221	.1971	.1130	.0606	.0302	.0139	.0059	1
2	.9998	.9988	.9963	.9917	.9848	.9752	.9630	.9481	.9305	.9104	.7788	.6174	.4552	.3127	.2001	.1189	.0652	.0327	2
3	1.000	1.000	.9998	.9993	.9984	.9970	.9947	.9915	.9871	.9815	.9306	.8389	.7133	.5696	.4256	.2963	.1911	.1133	3
4			1.000	1.000	.9999	.9997	.9995	.9990	.9983	.9972	.9841	.9496	.8854	.7897	.6683	.5328	.3971	.2744	4
5					1.000	1.000	1.000	.9999	.9998	.9997	.9973	.9883	.9657	.9218	.8513	.7535	.6331	.5000	5
6								1.000	1.000	1.000	.9997	.9980	.9924	.9784	.9499	.9006	.8262	.7256	6
7											1.000	.9998	.9988	.9957	.9878	.9707	.9390	.8867	7
8												1.000	.9999	.9994	.9980	.9941	.9852	.9673	8
9													1.000	1.000	.9998	.9993	.9978	.9941	9
10															1.000	1.000	.9998	.9995	10
11																	1.000	1.000	11
n = 12 0	.8864	.7847	.6938	.6127	.5404	.4759	.4186	.3677	.3225	.2824	.1422	.0687	.0317	.0138	.0057	.0022	.0008	.0002	0
1	.9938	.9769	.9514	.9191	.8816	.8405	.7967	.7513	.7052	.6590	.4435	.2749	.1584	.0850	.0424	.0196	.0083	.0032	1
2	.9998	.9985	.9952	.9893	.9804	.9684	.9532	.9348	.9134	.8891	.7358	.5583	.3907	.2528	.1513	.0834	.0421	.0193	2
3	1.000	.9999	.9997	.9990	.9978	.9957	.9925	.9880	.9820	.9744	.9078	.7946	.6488	.4925	.3467	.2253	.1345	.0730	3
4		1.000	1.000	.9999	.9998	.9996	.9991	.9984	.9973	.9957	.9761	.9274	.8424	.7237	.5833	.4382	.3044	.1938	4
5				1.000	1.000	1.000	.9999	.9998	.9997	.9995	.9954	.9806	.9456	.8822	.7873	.6652	.5269	.3872	5
6							1.000	1.000	1.000	.9999	.9993	.9961	.9857	.9614	.9154	.8418	.7393	.6128	6
7										1.000	.9999	.9994	.9972	.9905	.9745	.9427	.8883	.8062	7
8											1.000	.9999	.9996	.9983	.9944	.9847	.9644	.9270	8
9												1.000	1.000	.9998	.9992	.9972	.9921	.9807	9
10														1.000	.9999	.9997	.9989	.9968	10
11															1.000	1.000	.9999	.9998	11
12																	1.000	1.000	12
n = 13 0	.8775	.7690	.6730	.5882	.5133	.4474	.3893	.3383	.2935	.2542	.1209	.0550	.0238	.0097	.0037	.0013	.0004	.0001	0
1	.9928	.9730	.9436	.9068	.8646	.8186	.7702	.7206	.6707	.6213	.3983	.2336	.1267	.0637	.0296	.0126	.0049	.0017	1
2	.9997	.9980	.9938	.9865	.9755	.9608	.9422	.9201	.8946	.8661	.6920	.5017	.3326	.2025	.1132	.0579	.0269	.0112	2
3	1.000	.9999	.9995	.9986	.9969	.9940	.9897	.9837	.9758	.9658	.8820	.7473	.5843	.4206	.2783	.1686	.0929	.0461	3
4		1.000	1.000	.9999	.9997	.9993	.9987	.9976	.9959	.9935	.9658	.9009	.7940	.6543	.5005	.3530	.2279	.1334	4
5				1.000	1.000	.9999	.9999	.9997	.9995	.9991	.9925	.9700	.9198	.8346	.7159	.5744	.4268	.2905	5
6						1.000	1.000	1.000	.9999	.9999	.9987	.9930	.9757	.9376	.8705	.7712	.6437	.5000	6
7									1.000	1.000	.9998	.9988	.9944	.9818	.9538	.9023	.8212	.7095	7
8											1.000	.9998	.9990	.9960	.9874	.9679	.9302	.8666	8
9												1.000	.9999	.9993	.9975	.9922	.9797	.9539	9
10													1.000	.9999	.9997	.9987	.9959	.9888	10
11														1.000	1.000	.9999	.9995	.9983	11
12																1.000	1.000	.9999	12
13																		1.000	13

BINOMIAL DISTRIBUTION FUNCTION

x \ p	0.01	0.02	0.03	0.04	0.05	0.06	0.07	0.08	0.09	0.10	0.15	0.20	0.25	0.30	0.35	0.40	0.45	0.50	p / x
n = 14 0	.8687	.7536	.6528	.5647	.4877	.4205	.3620	.3112	.2670	.2288	.1028	.0440	.0178	.0068	.0024	.0008	.0002	.0001	0
1	.9916	.9690	.9355	.8941	.8470	.7963	.7436	.6900	.6368	.5846	.3567	.1979	.1010	.0475	.0205	.0081	.0029	.0009	1
2	.9997	.9975	.9923	.9833	.9699	.9522	.9302	.9042	.8745	.8416	.6479	.4481	.2811	.1608	.0839	.0398	.0170	.0065	2
3	1.000	.9999	.9994	.9981	.9958	.9920	.9864	.9786	.9685	.9559	.8535	.6982	.5213	.3552	.2205	.1243	.0632	.0287	3
4		1.000	1.000	.9998	.9996	.9990	.9980	.9965	.9941	.9908	.9533	.8702	.7415	.5842	.4227	.2793	.1672	.0898	4
5				1.000	1.000	.9999	.9998	.9996	.9992	.9985	.9885	.9561	.8883	.7805	.6405	.4859	.3373	.2120	5
6						1.000	1.000	1.000	.9999	.9998	.9978	.9884	.9617	.9067	.8164	.6925	.5461	.3953	6
7									1.000	1.000	.9997	.9976	.9897	.9685	.9247	.8499	.7414	.6047	7
8											1.000	.9996	.9978	.9917	.9757	.9417	.8811	.7880	8
9												1.000	.9997	.9983	.9940	.9825	.9574	.9102	9
10													1.000	.9998	.9989	.9961	.9886	.9713	10
11														1.000	.9999	.9994	.9978	.9935	11
12															1.000	.9999	.9997	.9991	12
13																1.000	1.000	.9999	13
14																		1.000	14
n = 15 0	.8601	.7386	.6333	.5421	.4633	.3953	.3367	.2863	.2430	.2059	.0874	.0352	.0134	.0047	.0016	.0005	.0001	.0000	0
1	.9904	.9647	.9270	.8809	.8290	.7738	.7168	.6597	.6035	.5490	.3186	.1671	.0802	.0353	.0142	.0052	.0017	.0005	1
2	.9996	.9970	.9906	.9797	.9638	.9429	.9171	.8870	.8531	.8159	.6042	.3980	.2361	.1268	.0617	.0271	.0107	.0037	2
3	1.000	.9998	.9992	.9976	.9945	.9896	.9825	.9727	.9601	.9444	.8227	.6482	.4613	.2969	.1727	.0905	.0424	.0176	3
4		1.000	.9999	.9998	.9994	.9986	.9972	.9950	.9918	.9873	.9383	.8358	.6865	.5155	.3519	.2173	.1204	.0592	4
5			1.000	1.000	.9999	.9999	.9997	.9993	.9987	.9978	.9832	.9389	.8516	.7216	.5643	.4032	.2608	.1509	5
6					1.000	1.000	1.000	.9999	.9998	.9997	.9964	.9819	.9434	.8689	.7548	.6098	.4522	.3036	6
7								1.000	1.000	1.000	.9994	.9958	.9827	.9500	.8868	.7869	.6535	.5000	7
8											.9999	.9992	.9958	.9848	.9578	.9050	.8182	.6964	8
9											1.000	.9999	.9992	.9963	.9876	.9662	.9231	.8491	9
10												1.000	.9999	.9993	.9972	.9907	.9745	.9408	10
11													1.000	.9999	.9995	.9981	.9937	.9824	11
12														1.000	.9999	.9997	.9989	.9963	12
13															1.000	1.000	.9999	.9995	13
14																	1.000	1.000	14
n = 20 0	.8179	.6676	.5438	.4420	.3585	.2901	.2342	.1887	.1516	.1216	.0388	.0115	.0032	.0008	.0002	.0000	.0000	.0000	0
1	.9831	.9401	.8802	.8103	.7358	.6605	.5869	.5169	.4516	.3917	.1756	.0692	.0243	.0076	.0021	.0005	.0001	.0000	1
2	.9990	.9929	.9790	.9561	.9245	.8850	.8390	.7879	.7334	.6769	.4049	.2061	.0913	.0355	.0121	.0036	.0009	.0002	2
3	1.000	.9994	.9973	.9926	.9841	.9710	.9529	.9294	.9007	.8670	.6477	.4114	.2252	.1071	.0444	.0160	.0049	.0013	3
4		1.000	.9997	.9990	.9974	.9944	.9893	.9817	.9710	.9568	.8298	.6296	.4148	.2375	.1182	.0510	.0189	.0059	4
5			1.000	.9999	.9997	.9991	.9981	.9962	.9932	.9887	.9327	.8042	.6172	.4164	.2454	.1256	.0553	.0207	5
6				1.000	1.000	.9999	.9997	.9994	.9987	.9976	.9781	.9133	.7858	.6080	.4166	.2500	.1299	.0577	6
7						1.000	1.000	.9999	.9998	.9996	.9941	.9679	.8982	.7723	.6010	.4159	.2520	.1316	7
8								1.000	1.000	.9999	.9987	.9900	.9591	.8867	.7624	.5956	.4143	.2517	8
9										1.000	.9998	.9974	.9861	.9520	.8782	.7553	.5914	.4119	9
10											1.000	.9994	.9961	.9829	.9468	.8725	.7507	.5881	10
11												.9999	.9991	.9949	.9804	.9435	.8692	.7483	11
12												1.000	.9998	.9987	.9940	.9790	.9420	.8684	12
13													1.000	.9997	.9985	.9935	.9786	.9423	13
14														1.000	.9997	.9984	.9936	.9793	14
15															1.000	.9997	.9985	.9941	15
16																1.000	.9997	.9987	16
17																	1.000	.9998	17
18																		1.000	18

BINOMIAL DISTRIBUTION FUNCTION

	x \ p	0.01	0.02	0.03	0.04	0.05	0.06	0.07	0.08	0.09	0.10	0.15	0.20	0.25	0.30	0.35	0.40	0.45	0.50	p / x
n = 25	0	.7778	.6035	.4670	.3604	.2774	.2129	.1630	.1244	.0946	.0718	.0172	.0038	.0008	.0001	.0000	.0000	.0000	.0000	0
	1	.9742	.9114	.8280	.7358	.6424	.5527	.4696	.3947	.3286	.2712	.0931	.0274	.0070	.0016	.0003	.0001	.0000	.0000	1
	2	.9980	.9868	.9620	.9235	.8729	.8129	.7466	.6768	.6063	.5371	.2537	.0982	.0321	.0090	.0021	.0004	.0001	.0000	2
	3	.9999	.9986	.9938	.9835	.9659	.9402	.9064	.8649	.8169	.7636	.4711	.2340	.0962	.0332	.0097	.0024	.0005	.0001	3
	4	1.000	.9999	.9992	.9972	.9928	.9850	.9726	.9549	.9314	.9020	.6821	.4207	.2137	.0905	.0320	.0095	.0023	.0005	4
	5		1.000	.9999	.9996	.9988	.9969	.9935	.9877	.9790	.9666	.8385	.6167	.3783	.1935	.0826	.0294	.0086	.0020	5
	6			1.000	1.000	.9998	.9995	.9987	.9972	.9946	.9905	.9305	.7800	.5611	.3407	.1734	.0736	.0258	.0073	6
	7					1.000	.9999	.9998	.9995	.9989	.9977	.9745	.8909	.7265	.5118	.3061	.1536	.0639	.0216	7
	8						1.000	1.000	.9999	.9998	.9995	.9920	.9532	.8506	.6769	.4668	.2735	.1340	.0539	8
	9								1.000	1.000	.9999	.9979	.9827	.9287	.8106	.6303	.4246	.2424	.1148	9
	10										1.000	.9995	.9944	.9703	.9022	.7712	.5858	.3843	.2122	10
	11											.9999	.9985	.9893	.9558	.8746	.7323	.5426	.3450	11
	12											1.000	.9996	.9966	.9825	.9396	.8462	.6937	.5000	12
	13												.9999	.9991	.9940	.9745	.9222	.8173	.6550	13
	14												1.000	.9998	.9982	.9907	.9656	.9040	.7878	14
	15													1.000	.9995	.9971	.9868	.9560	.8852	15
	16														.9999	.9992	.9957	.9826	.9461	16
	17														1.000	.9998	.9988	.9942	.9784	17
	18															1.000	.9997	.9984	.9927	18
	19																.9999	.9996	.9980	19
	20																1.000	.9999	.9995	20
	21																	1.000	.9999	21
	22																		1.000	22
n = 30	0	.7397	.5455	.4010	.2939	.2146	.1563	.1134	.0820	.0591	.0424	.0076	.0012	.0002	.0000	.0000	.0000	.0000	.0000	0
	1	.9639	.8795	.7731	.6612	.5535	.4555	.3694	.2958	.2343	.1837	.0480	.0105	.0020	.0003	.0000	.0000	.0000	.0000	1
	2	.9967	.9783	.9399	.8831	.8122	.7324	.6487	.5654	.4855	.4114	.1514	.0442	.0106	.0021	.0003	.0000	.0000	.0000	2
	3	.9998	.9971	.9881	.9694	.9392	.8974	.8450	.7842	.7175	.6474	.3217	.1227	.0374	.0093	.0019	.0003	.0000	.0000	3
	4	1.000	.9997	.9982	.9937	.9844	.9685	.9447	.9126	.8723	.8245	.5245	.2552	.0979	.0302	.0075	.0015	.0002	.0000	4
	5		1.000	.9998	.9989	.9967	.9921	.9838	.9707	.9519	.9268	.7106	.4275	.2026	.0766	.0233	.0057	.0011	.0002	5
	6			1.000	.9999	.9994	.9983	.9960	.9918	.9848	.9742	.8474	.6070	.3481	.1595	.0586	.0172	.0040	.0007	6
	7				1.000	.9999	.9997	.9992	.9980	.9959	.9922	.9302	.7608	.5143	.2814	.1238	.0435	.0121	.0026	7
	8					1.000	1.000	.9999	.9996	.9990	.9980	.9722	.8713	.6736	.4315	.2247	.0940	.0312	.0081	8
	9							1.000	.9999	.9998	.9995	.9903	.9389	.8034	.5888	.3575	.1763	.0694	.0214	9
	10								1.000	1.000	.9999	.9971	.9744	.8943	.7304	.5078	.2915	.1350	.0494	10
	11										1.000	.9992	.9905	.9493	.8407	.6548	.4311	.2327	.1002	11
	12											.9998	.9969	.9784	.9155	.7802	.5785	.3592	.1808	12
	13											1.000	.9991	.9918	.9599	.8737	.7145	.5025	.2923	13
	14												.9998	.9973	.9831	.9348	.8246	.6448	.4278	14
	15												.9999	.9992	.9936	.9699	.9029	.7691	.5722	15
	16												1.000	.9998	.9979	.9876	.9519	.8644	.7077	16
	17													.9999	.9994	.9955	.9788	.9286	.8192	17
	18													1.000	.9998	.9986	.9917	.9666	.8998	18
	19														1.000	.9996	.9971	.9862	.9506	19
	20															.9999	.9991	.9950	.9786	20
	21															1.000	.9998	.9984	.9919	21
	22																1.000	.9996	.9974	22
	23																	.9999	.9993	23
	24																	1.000	.9998	24
	25																		1.000	25

BINOMIAL DISTRIBUTION FUNCTION

x \ p	0.01	0.02	0.03	0.04	0.05	0.06	0.07	0.08	0.09	0.10	0.15	0.20	0.25	0.30	0.35	0.40	0.45	0.50	p / x
n = 40 0	.6690	.4457	.2957	.1954	.1285	.0842	.0549	.0356	.0230	.0148	.0015	.0001	.0000	.0000	.0000	.0000	.0000	.0000	0
1	.9393	.8095	.6615	.5210	.3991	.2990	.2201	.1594	.1140	.0805	.0121	.0015	.0001	.0000	.0000	.0000	.0000	.0000	1
2	.9925	.9543	.8822	.7855	.6767	.5665	.4625	.3694	.2894	.2228	.0486	.0079	.0010	.0001	.0000	.0000	.0000	.0000	2
3	.9993	.9918	.9686	.9252	.8619	.7827	.6937	.6007	.5092	.4231	.1302	.0285	.0047	.0006	.0001	.0000	.0000	.0000	3
4	1.000	.9988	.9933	.9790	.9520	.9104	.8546	.7868	.7103	.6290	.2633	.0759	.0160	.0026	.0003	.0000	.0000	.0000	4
5		.9999	.9988	.9951	.9861	.9691	.9419	.9033	.8535	.7937	.4325	.1613	.0433	.0086	.0013	.0001	.0000	.0000	5
6		1.000	.9998	.9990	.9966	.9909	.9801	.9624	.9361	.9005	.6067	.2859	.0962	.0238	.0044	.0006	.0001	.0000	6
7			1.000	.9998	.9993	.9977	.9942	.9873	.9758	.9581	.7559	.4371	.1820	.0553	.0124	.0021	.0002	.0000	7
8				1.000	.9999	.9995	.9985	.9963	.9919	.9845	.8646	.5931	.2998	.1110	.0303	.0061	.0009	.0001	8
9					1.000	.9999	.9997	.9990	.9976	.9949	.9328	.7318	.4395	.1959	.0644	.0156	.0027	.0003	9
10						1.000	.9999	.9998	.9994	.9985	.9701	.8392	.5839	.3087	.1215	.0352	.0074	.0011	10
11							1.000	1.000	.9999	.9996	.9880	.9125	.7151	.4406	.2053	.0709	.0179	.0032	11
12									1.000	.9999	.9957	.9568	.8209	.5772	.3143	.1285	.0386	.0083	12
13										1.000	.9986	.9806	.8968	.7032	.4408	.2112	.0751	.0192	13
14											.9996	.9921	.9456	.8074	.5721	.3174	.1326	.0403	14
15											.9999	.9971	.9738	.8849	.6946	.4402	.2142	.0769	15
16											1.000	.9990	.9884	.9367	.7978	.5681	.3185	.1341	16
17												.9997	.9953	.9680	.8761	.6885	.4391	.2148	17
18												.9999	.9983	.9852	.9301	.7911	.5651	.3179	18
19												1.000	.9994	.9937	.9637	.8702	.6844	.4373	19
20													.9998	.9976	.9827	.9256	.7870	.5627	20
21													1.000	.9991	.9925	.9608	.8669	.6821	21
22														.9997	.9970	.9811	.9233	.7852	22
23														.9999	.9989	.9917	.9595	.8659	23
24														1.000	.9996	.9966	.9804	.9231	24
25															.9999	.9988	.9914	.9597	25
26															1.000	.9996	.9966	.9808	26
27																.9999	.9988	.9917	27
28																1.000	.9996	.9968	28
29																	.9999	.9989	29
30																	1.000	.9997	30
31																		.9999	31
32																		1.000	32
n = 50 0	.6050	.3642	.2181	.1299	.0769	.0453	.0266	.0155	.0090	.0052	.0003	.0000	.0000	.0000	.0000	.0000	.0000	.0000	0
1	.9106	.7358	.5553	.4005	.2794	.1900	.1265	.0827	.0532	.0338	.0029	.0002	.0000	.0000	.0000	.0000	.0000	.0000	1
2	.9862	.9216	.8108	.6767	.5405	.4162	.3108	.2260	.1605	.1117	.0142	.0013	.0001	.0000	.0000	.0000	.0000	.0000	2
3	.9984	.9822	.9372	.8609	.7604	.6473	.5327	.4253	.3303	.2503	.0460	.0057	.0005	.0000	.0000	.0000	.0000	.0000	3
4	.9999	.9968	.9832	.9510	.8964	.8206	.7290	.6290	.5277	.4312	.1121	.0185	.0021	.0002	.0000	.0000	.0000	.0000	4
5	1.000	.9995	.9963	.9856	.9622	.9224	.8650	.7919	.7072	.6161	.2194	.0480	.0070	.0007	.0001	.0000	.0000	.0000	5
6		.9999	.9993	.9964	.9882	.9711	.9417	.8981	.8404	.7702	.3613	.1034	.0194	.0025	.0002	.0000	.0000	.0000	6
7		1.000	.9999	.9992	.9968	.9906	.9780	.9562	.9232	.8779	.5188	.1904	.0453	.0073	.0008	.0001	.0000	.0000	7
8			1.000	.9999	.9992	.9973	.9927	.9833	.9672	.9421	.6681	.3073	.0916	.0183	.0025	.0002	.0000	.0000	8
9				1.000	.9998	.9993	.9978	.9944	.9875	.9755	.7911	.4437	.1637	.0402	.0067	.0008	.0001	.0000	9
10					1.000	.9998	.9994	.9983	.9957	.9906	.8801	.5836	.2622	.0789	.0160	.0022	.0002	.0000	10
11						1.000	.9999	.9995	.9987	.9968	.9372	.7107	.3816	.1390	.0342	.0057	.0006	.0000	11
12							1.000	.9999	.9996	.9990	.9699	.8139	.5110	.2229	.0661	.0133	.0018	.0002	12
13								1.000	.9999	.9997	.9868	.8894	.6370	.3279	.1163	.0280	.0045	.0005	13
14									1.000	.9999	.9947	.9393	.7481	.4468	.1878	.0540	.0104	.0013	14
15										1.000	.9981	.9692	.8369	.5692	.2801	.0955	.0220	.0033	15
16											.9993	.9856	.9017	.6839	.3889	.1561	.0427	.0077	16
17											.9998	.9937	.9449	.7822	.5060	.2369	.0765	.0164	17
18											.9999	.9975	.9713	.8594	.6216	.3356	.1273	.0325	18
19											1.000	.9991	.9861	.9152	.7264	.4465	.1974	.0595	19
20												.9997	.9937	.9522	.8139	.5610	.2862	.1013	20
21												.9999	.9974	.9749	.8813	.6701	.3900	.1611	21
22												1.000	.9990	.9877	.9290	.7660	.5019	.2399	22
23													.9996	.9944	.9604	.8438	.6134	.3359	23
24													.9999	.9976	.9793	.9022	.7160	.4439	24
25													1.000	.9991	.9900	.9427	.8034	.5561	25
26														.9997	.9955	.9686	.8721	.6641	26
27														.9999	.9981	.9840	.9220	.7601	27
28														1.000	.9993	.9924	.9556	.8389	28
29															.9997	.9966	.9765	.8987	29
30															.9999	.9986	.9884	.9405	30
31															1.000	.9995	.9947	.9675	31
32																.9998	.9978	.9836	32
33																.9999	.9991	.9923	33
34																1.000	.9997	.9967	34
35																	.9999	.9987	35
36																	1.000	.9995	36
37																		.9998	37
38																		1.000	38

TABLE 2 POISSON DISTRIBUTION FUNCTION

The table gives the probability that a Poisson random variable X with mean m is less than or equal to x, i.e.

$$P(X \leq x) = \sum_{r=0}^{x} m^r \frac{e^{-m}}{r!}$$

x \ m	0.1	0.2	0.3	0.4	0.5	0.6	0.7	0.8	0.9	1.0	1.2	1.4	1.6	1.8	m / x
0	.9048	.8187	.7408	.6703	.6065	.5488	.4966	.4493	.4066	.3679	.3012	.2466	.2019	.1653	0
1	.9953	.9825	.9631	.9384	.9098	.8781	.8442	.8088	.7725	.7358	.6626	.5918	.5249	.4628	1
2	.9998	.9989	.9964	.9921	.9856	.9769	.9659	.9526	.9371	.9197	.8795	.8335	.7834	.7306	2
3	1.000	.9999	.9997	.9992	.9982	.9966	.9942	.9909	.9865	.9810	.9662	.9463	.9212	.8913	3
4		1.000	1.000	.9999	.9998	.9996	.9992	.9986	.9977	.9963	.9923	.9857	.9763	.9636	4
5				1.000	1.000	1.000	.9999	.9998	.9997	.9994	.9985	.9968	.9940	.9896	5
6							1.000	1.000	1.000	.9999	.9997	.9994	.9987	.9974	6
7										1.000	1.000	.9999	.9997	.9994	7
8												1.000	1.000	.9999	8
9														1.000	9

x \ m	2.0	2.2	2.4	2.6	2.8	3.0	3.2	3.4	3.6	3.8	4.0	4.5	5.0	5.5	m / x
0	.1353	.1108	.0907	.0743	.0608	.0498	.0408	.0334	.0273	.0224	.0183	.0111	.0067	.0041	0
1	.4060	.3546	.3084	.2674	.2311	.1991	.1712	.1468	.1257	.1074	.0916	.0611	.0404	.0266	1
2	.6767	.6227	.5697	.5184	.4695	.4232	.3799	.3397	.3027	.2689	.2381	.1736	.1247	.0884	2
3	.8571	.8194	.7787	.7360	.6919	.6472	.6025	.5584	.5152	.4735	.4335	.3423	.2650	.2017	3
4	.9473	.9275	.9041	.8774	.8477	.8153	.7806	.7442	.7064	.6678	.6288	.5321	.4405	.3575	4
5	.9834	.9751	.9643	.9510	.9349	.9161	.8946	.8705	.8441	.8156	.7851	.7029	.6160	.5289	5
6	.9955	.9925	.9884	.9828	.9756	.9665	.9554	.9421	.9267	.9091	.8893	.8311	.7622	.6860	6
7	.9989	.9980	.9967	.9947	.9919	.9881	.9832	.9769	.9692	.9599	.9489	.9134	.8666	.8095	7
8	.9998	.9995	.9991	.9985	.9976	.9962	.9943	.9917	.9883	.9840	.9786	.9597	.9319	.8944	8
9	1.000	.9999	.9998	.9996	.9993	.9989	.9982	.9973	.9960	.9942	.9919	.9829	.9682	.9462	9
10		1.000	1.000	.9999	.9998	.9997	.9995	.9992	.9987	.9981	.9972	.9933	.9863	.9747	10
11				1.000	1.000	.9999	.9999	.9998	.9996	.9994	.9991	.9976	.9945	.9890	11
12						1.000	1.000	.9999	.9999	.9998	.9997	.9992	.9980	.9955	12
13								1.000	1.000	1.000	.9999	.9997	.9993	.9983	13
14											1.000	.9999	.9998	.9994	14
15												1.000	.9999	.9998	15
16													1.000	.9999	16
17														1.000	17

x \ m	6.0	6.5	7.0	7.5	8.0	8.5	9.0	9.5	10.0	11.0	12.0	13.0	14.0	15.0	m / x
0	.0025	.0015	.0009	.0006	.0003	.0002	.0001	.0001	.0000	.0000	.0000	.0000	.0000	.0000	0
1	.0174	.0113	.0073	.0047	.0030	.0019	.0012	.0008	.0005	.0002	.0001	.0000	.0000	.0000	1
2	.0620	.0430	.0296	.0203	.0138	.0093	.0062	.0042	.0028	.0012	.0005	.0002	.0001	.0000	2
3	.1512	.1118	.0818	.0591	.0424	.0301	.0212	.0149	.0103	.0049	.0023	.0011	.0005	.0002	3
4	.2851	.2237	.1730	.1321	.0996	.0744	.0550	.0403	.0293	.0151	.0076	.0037	.0018	.0009	4
5	.4457	.3690	.3007	.2414	.1912	.1496	.1157	.0885	.0671	.0375	.0203	.0107	.0055	.0028	5
6	.6063	.5265	.4497	.3782	.3134	.2562	.2068	.1649	.1301	.0786	.0458	.0259	.0142	.0076	6
7	.7440	.6728	.5987	.5246	.4530	.3856	.3239	.2687	.2202	.1432	.0895	.0540	.0316	.0180	7
8	.8472	.7916	.7291	.6620	.5925	.5231	.4557	.3918	.3328	.2320	.1550	.0998	.0621	.0374	8
9	.9161	.8774	.8305	.7764	.7166	.6530	.5874	.5218	.4579	.3405	.2424	.1658	.1094	.0699	9
10	.9574	.9332	.9015	.8622	.8159	.7634	.7060	.6453	.5830	.4599	.3472	.2517	.1757	.1185	10
11	.9799	.9661	.9467	.9208	.8881	.8487	.8030	.7520	.6968	.5793	.4616	.3532	.2600	.1848	11
12	.9912	.9840	.9730	.9573	.9362	.9091	.8758	.8364	.7916	.6887	.5760	.4631	.3585	.2676	12
13	.9964	.9929	.9872	.9784	.9658	.9486	.9261	.8981	.8645	.7813	.6815	.5730	.4644	.3632	13
14	.9986	.9970	.9943	.9897	.9827	.9726	.9585	.9400	.9165	.8540	.7720	.6751	.5704	.4657	14
15	.9995	.9988	.9976	.9954	.9918	.9862	.9780	.9665	.9513	.9074	.8444	.7636	.6694	.5681	15
16	.9998	.9996	.9990	.9980	.9963	.9934	.9889	.9823	.9730	.9441	.8987	.8355	.7559	.6641	16
17	.9999	.9998	.9996	.9992	.9984	.9970	.9947	.9911	.9857	.9678	.9370	.8905	.8272	.7489	17
18	1.000	.9999	.9999	.9997	.9993	.9987	.9976	.9957	.9928	.9823	.9626	.9302	.8826	.8195	18
19		1.000	1.000	.9999	.9997	.9995	.9989	.9980	.9965	.9907	.9787	.9573	.9235	.8752	19
20				1.000	.9999	.9998	.9996	.9991	.9984	.9953	.9884	.9750	.9521	.9170	20
21					1.000	.9999	.9998	.9996	.9993	.9977	.9939	.9859	.9712	.9469	21
22						1.000	.9999	.9999	.9997	.9990	.9970	.9924	.9833	.9673	22
23							1.000	.9999	.9999	.9995	.9985	.9960	.9907	.9805	23
24								1.000	1.000	.9998	.9993	.9980	.9950	.9888	24
25										.9999	.9997	.9990	.9974	.9938	25
26										1.000	.9999	.9995	.9987	.9967	26
27											.9999	.9998	.9994	.9983	27
28											1.000	.9999	.9997	.9991	28
29												1.000	.9999	.9996	29
30													.9999	.9998	30
31													1.000	.9999	31
32														1.000	32

TABLE 3 NORMAL DISTRIBUTION FUNCTION

The table gives the probability p that a normally distributed random variable Z with zero mean and unit variance is less than or equal to z.

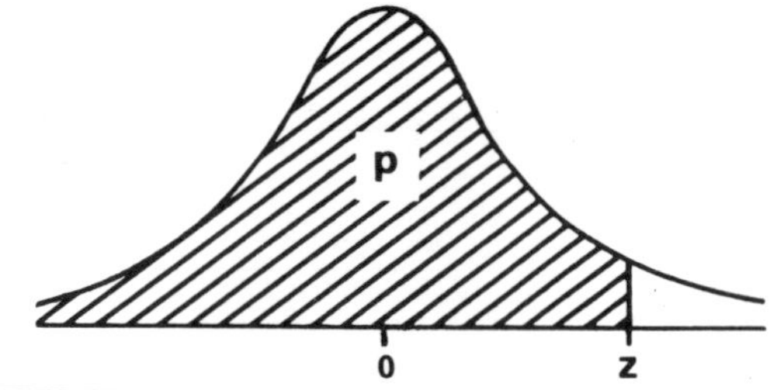

z	.00	.01	.02	.03	.04	.05	.06	.07	.08	.09
0.0	.50000	.50399	.50798	.51197	.51595	.51994	.52392	.52790	.53188	.53586
0.1	.53983	.54380	.54776	.55172	.55567	.55962	.56356	.56749	.57142	.57535
0.2	.57926	.58317	.58706	.59095	.59483	.59871	.60257	.60642	.61026	.61409
0.3	.61791	.62172	.62552	.62930	.63307	.63683	.64058	.64431	.64803	.65173
0.4	.65542	.65910	.66276	.66640	.67003	.67364	.67724	.68082	.68439	.68793
0.5	.69146	.69497	.69847	.70194	.70540	.70884	.71226	.71566	.71904	.72240
0.6	.72575	.72907	.73237	.73565	.73891	.74215	.74537	.74857	.75175	.75490
0.7	.75804	.76115	.76424	.76730	.77035	.77337	.77637	.77935	.78230	.78524
0.8	.78814	.79103	.79389	.79673	.79955	.80234	.80511	.80785	.81057	.81327
0.9	.81594	.81859	.82121	.82381	.82639	.82894	.83147	.83398	.83646	.83891
1.0	.84134	.84375	.84614	.84849	.85083	.85314	.85543	.85769	.85993	.86214
1.1	.86433	.86650	.86864	.87076	.87286	.87493	.87698	.87900	.88100	.88298
1.2	.88493	.88686	.88877	.89065	.89251	.89435	.89617	.89796	.89973	.90147
1.3	.90320	.90490	.90658	.90824	.90988	.91149	.91309	.91466	.91621	.91774
1.4	.91924	.92073	.92220	.92364	.92507	.92647	.92785	.92922	.93056	.93189
1.5	.93319	.93448	.93574	.93699	.93822	.93943	.94062	.94179	.94295	.94408
1.6	.94520	.94630	.94738	.94845	.94950	.95053	.95154	.95254	.95352	.95449
1.7	.95543	.95637	.95728	.95818	.95907	.95994	.96080	.96164	.96246	.96327
1.8	.96407	.96485	.96562	.96638	.96712	.96784	.96856	.96926	.96995	.97062
1.9	.97128	.97193	.97257	.97320	.97381	.97441	.97500	.97558	.97615	.97670
2.0	.97725	.97778	.97831	.97882	.97932	.97982	.98030	.98077	.98124	.98169
2.1	.98214	.98257	.98300	.98341	.98382	.98422	.98461	.98500	.98537	.98574
2.2	.98610	.98645	.98679	.98713	.98745	.98778	.98809	.98840	.98870	.98899
2.3	.98928	.98956	.98983	.99010	.99036	.99061	.99086	.99111	.99134	.99158
2.4	.99180	.99202	.99224	.99245	.99266	.99286	.99305	.99324	.99343	.99361
2.5	.99379	.99396	.99413	.99430	.99446	.99461	.99477	.99492	.99506	.99520
2.6	.99534	.99547	.99560	.99573	.99585	.99598	.99609	.99621	.99632	.99643
2.7	.99653	.99664	.99674	.99683	.99693	.99702	.99711	.99720	.99728	.99736
2.8	.99744	.99752	.99760	.99767	.99774	.99781	.99788	.99795	.99801	.99807
2.9	.99813	.99819	.99825	.99831	.99836	.99841	.99846	.99851	.99856	.99861
3.0	.99865	.99869	.99874	.99878	.99882	.99886	.99889	.99893	.99896	.99900
3.1	.99903	.99906	.99910	.99913	.99916	.99918	.99921	.99924	.99926	.99929
3.2	.99931	.99934	.99936	.99938	.99940	.99942	.99944	.99946	.99948	.99950
3.3	.99952	.99953	.99955	.99957	.99958	.99960	.99961	.99962	.99964	.99965
3.4	.99966	.99968	.99969	.99970	.99971	.99972	.99973	.99974	.99975	.99976
3.5	.99977	.99978	.99978	.99979	.99980	.99981	.99981	.99982	.99983	.99983
3.6	.99984	.99985	.99985	.99986	.99986	.99987	.99987	.99988	.99988	.99989
3.7	.99989	.99990	.99990	.99990	.99991	.99991	.99992	.99992	.99992	.99992
3.8	.99993	.99993	.99993	.99994	.99994	.99994	.99994	.99995	.99995	.99995
3.9	.99995	.99995	.99996	.99996	.99996	.99996	.99996	.99996	.99997	.99997

TABLE 4 PERCENTAGE POINTS OF THE NORMAL DISTRIBUTION

The table gives the values of z satisfying

$$P(Z \le z) = p$$

where Z is a normally distributed random variable with zero mean and unit variance.

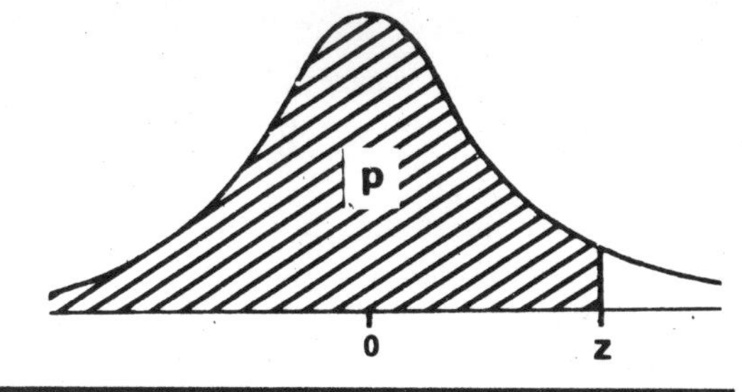

p	.00	.01	.02	.03	.04	.05	.06	.07	.08	.09
0.50	0.000	0.025	0.050	0.075	0.100	0.126	0.151	0.176	0.202	0.228
0.60	0.253	0.279	0.305	0.332	0.358	0.385	0.412	0.440	0.468	0.496
0.70	0.524	0.553	0.583	0.613	0.643	0.674	0.706	0.739	0.772	0.806
0.80	0.842	0.878	0.915	0.954	0.994	1.036	1.080	1.126	1.175	1.227
0.90	1.282	1.341	1.405	1.476	1.555					

p	.000	.001	.002	.003	.004	.005	.006	.007	.008	.009
0.95	1.645	1.655	1.665	1.675	1.685	1.695	1.706	1.717	1.728	1.739
0.96	1.751	1.762	1.774	1.787	1.799	1.812	1.825	1.838	1.852	1.866
0.97	1.881	1.896	1.911	1.927	1.943	1.960	1.977	1.995	2.014	2.034
0.98	2.054	2.075	2.097	2.120	2.144	2.170	2.197	2.226	2.257	2.290
0.99	2.326	2.366	2.409	2.457	2.512	2.576	2.652	2.748	2.878	3.090

TABLE 5 CONTROL CHART LIMITS FOR SAMPLE RANGE

The table gives

(i) values of k satisfying $\sigma = kE(W)$, where $E(W)$ may be estimated by W,

(ii) values of $D_{1-\alpha}$ satisfying $P(W \le D_{1-\alpha}\sigma) = 1 - \alpha$,

(iii) values of $D'_{1-\alpha}$ satisfying $P(W \le D'_{1-\alpha}E(W)) = 1 - \alpha$

where W is the range of a random sample of size n from a normal distribution with standard deviation σ.

n	k	$D_{.975}$	$D_{.999}$	$D'_{.975}$	$D'_{.999}$
2	0.886	3.170	4.654	2.809	4.124
3	0.591	3.682	5.064	2.176	2.992
4	0.486	3.984	5.309	1.935	2.579
5	0.430	4.197	5.484	1.804	2.358
6	0.395	4.361	5.619	1.721	2.217
7	0.370	4.493	5.729	1.662	2.119
8	0.351	4.605	5.823	1.617	2.045
9	0.337	4.700	5.903	1.583	1.988
10	0.325	4.785	5.974	1.555	1.941

TABLE 6 PERCENTAGE POINTS OF THE χ^2- DISTRIBUTION

The table gives the values of x satisfying

$$P(X \le x) = p$$

where X is a χ^2 random variable with υ degrees of freedom.

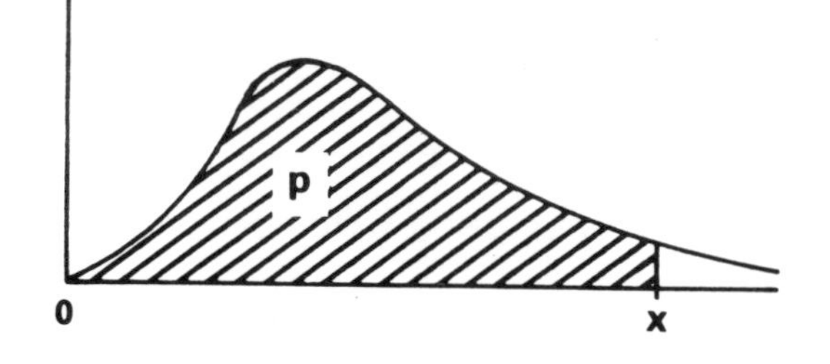

υ \ p	0.005	0.01	0.025	0.05	0.1	0.9	0.95	0.975	0.99	0.995
1	0.00004	0.0002	0.001	0.004	0.016	2.706	3.841	5.024	6.635	7.879
2	0.010	0.020	0.051	0.103	0.211	4.605	5.991	7.378	9.210	10.597
3	0.072	0.115	0.216	0.352	0.584	6.251	7.815	9.348	11.345	12.838
4	0.207	0.297	0.484	0.711	1.064	7.779	9.488	11.143	13.277	14.860
5	0.412	0.554	0.831	1.145	1.610	9.236	11.070	12.833	15.086	16.750
6	0.676	0.872	1.237	1.635	2.204	10.645	12.592	14.449	16.812	18.548
7	0.989	1.239	1.690	2.167	2.833	12.017	14.067	16.013	18.475	20.278
8	1.344	1.646	2.180	2.733	3.490	13.362	15.507	17.535	20.090	21.955
9	1.735	2.088	2.700	3.325	4.168	14.684	16.919	19.023	21.666	23.589
10	2.156	2.558	3.247	3.940	4.865	15.987	18.307	20.483	23.209	25.188
11	2.603	3.053	3.816	4.575	5.578	17.275	19.675	21.920	24.725	26.757
12	3.074	3.571	4.404	5.226	6.304	18.549	21.026	23.337	26.217	28.300
13	3.565	4.107	5.009	5.892	7.042	19.812	22.362	24.736	27.688	29.819
14	4.075	4.660	5.629	6.571	7.790	21.064	23.685	26.119	29.141	31.319
15	4.601	5.229	6.262	7.261	8.547	22.307	24.996	27.488	30.578	32.801
16	5.142	5.812	6.908	7.962	9.312	23.542	26.296	28.845	32.000	34.267
17	5.697	6.408	7.564	8.672	10.085	24.769	27.587	30.191	33.409	35.718
18	6.265	7.015	8.231	9.390	10.865	25.989	28.869	31.526	34.805	37.156
19	6.844	7.633	8.907	10.117	11.651	27.204	30.144	32.852	36.191	38.582
20	7.434	8.260	9.591	10.851	12.443	28.412	31.410	34.170	37.566	39.997
21	8.034	8.897	10.283	11.591	13.240	29.615	32.671	35.479	38.932	41.401
22	8.643	9.542	10.982	12.338	14.041	30.813	33.924	36.781	40.289	42.796
23	9.260	10.196	11.689	13.091	14.848	32.007	35.172	38.076	41.638	44.181
24	9.886	10.856	12.401	13.848	15.659	33.196	36.415	39.364	42.980	45.559
25	10.520	11.524	13.120	14.611	16.473	34.382	37.652	40.646	44.314	46.928
26	11.160	12.198	13.844	15.379	17.292	35.563	38.885	41.923	45.642	48.290
27	11.808	12.879	14.573	16.151	18.114	36.741	40.113	43.195	46.963	49.645
28	12.461	13.565	15.308	16.928	18.939	37.916	41.337	44.461	48.278	50.993
29	13.121	14.256	16.047	17.708	19.768	39.087	42.557	45.722	49.588	52.336
30	13.787	14.953	16.791	18.493	20.599	40.256	43.773	46.979	50.892	53.672
31	14.458	15.655	17.539	19.281	21.434	41.422	44.985	48.232	52.191	55.003
32	15.134	16.362	18.291	20.072	22.271	42.585	46.194	49.480	53.486	56.328
33	15.815	17.074	19.047	20.867	23.110	43.745	47.400	50.725	54.776	57.648
34	16.501	17.789	19.806	21.664	23.952	44.903	48.602	51.966	56.061	58.964
35	17.192	18.509	20.569	22.465	24.797	46.059	49.802	53.203	57.342	60.275
36	17.887	19.233	21.336	23.269	25.643	47.212	50.998	54.437	58.619	61.581
37	18.586	19.960	22.106	24.075	26.492	48.363	52.192	55.668	59.892	62.883
38	19.289	20.691	22.878	24.884	27.343	49.513	53.384	56.896	61.162	64.181
39	19.996	21.426	23.654	25.695	28.196	50.660	54.572	58.120	62.428	65.476
40	20.707	22.164	24.433	26.509	29.051	51.805	55.758	59.342	63.691	66.766
45	24.311	25.901	28.366	30.612	33.350	57.505	61.656	65.410	69.957	73.166
50	27.991	29.707	32.357	34.764	37.689	63.167	67.505	71.420	76.154	79.490
55	31.735	33.570	36.398	38.958	42.060	68.796	73.311	77.380	82.292	85.749
60	35.534	37.485	40.482	43.188	46.459	74.397	79.082	83.298	88.379	91.952
65	39.383	41.444	44.603	47.450	50.883	79.973	84.821	89.177	94.422	98.105
70	43.275	45.442	48.758	51.739	55.329	85.527	90.531	95.023	100.425	104.215
75	47.206	49.475	52.942	56.054	59.795	91.061	96.217	100.839	106.393	110.286
80	51.172	53.540	57.153	60.391	64.278	96.578	101.879	106.629	112.329	116.321
85	55.170	57.634	61.389	64.749	68.777	102.079	107.522	112.393	118.236	122.325
90	59.196	61.754	65.647	69.126	73.291	107.565	113.145	118.136	124.116	128.299
95	63.250	65.898	69.925	73.520	77.818	113.038	118.752	123.858	129.973	134.247
100	67.328	70.065	74.222	77.929	82.358	118.498	124.342	129.561	135.807	140.169

TABLE 7 PERCENTAGE POINTS OF THE STUDENT'S t-DISTRIBUTION

The table gives the values of x satisfying

$$P(X \leq x) = p$$

where X is a random variable having the Student's t-distribution with υ degrees of freedom.

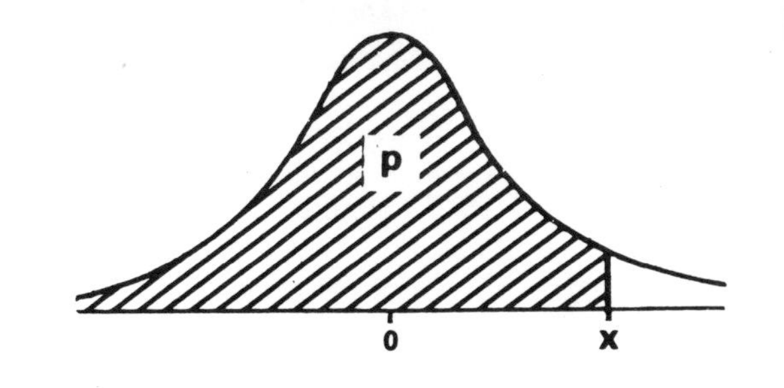

υ \ p	0.9	0.95	0.975	0.99	0.995
1	3.078	6.314	12.706	31.821	63.657
2	1.886	2.920	4.303	6.965	9.925
3	1.638	2.353	3.182	4.541	5.841
4	1.533	2.132	2.776	3.747	4.604
5	1.476	2.015	2.571	3.365	4.032
6	1.440	1.943	2.447	3.143	3.707
7	1.415	1.895	2.365	2.998	3.499
8	1.397	1.860	2.306	2.896	3.355
9	1.383	1.833	2.262	2.821	3.250
10	1.372	1.812	2.228	2.764	3.169
11	1.363	1.796	2.201	2.718	3.106
12	1.356	1.782	2.179	2.681	3.055
13	1.350	1.771	2.160	2.650	3.012
14	1.345	1.761	2.145	2.624	2.977
15	1.341	1.753	2.131	2.602	2.947
16	1.337	1.746	2.120	2.583	2.921
17	1.333	1.740	2.110	2.567	2.898
18	1.330	1.734	2.101	2.552	2.878
19	1.328	1.729	2.093	2.539	2.861
20	1.325	1.725	2.086	2.528	2.845
21	1.323	1.721	2.080	2.518	2.831
22	1.321	1.717	2.074	2.508	2.819
23	1.319	1.714	2.069	2.500	2.807
24	1.318	1.711	2.064	2.492	2.797
25	1.316	1.708	2.060	2.485	2.787
26	1.315	1.706	2.056	2.479	2.779
27	1.314	1.703	2.052	2.473	2.771
28	1.313	1.701	2.048	2.467	2.763

υ \ p	0.9	0.95	0.975	0.99	0.995
29	1.311	1.699	2.045	2.462	2.756
30	1.310	1.697	2.042	2.457	2.750
31	1.309	1.696	2.040	2.453	2.744
32	1.309	1.694	2.037	2.449	2.738
33	1.308	1.692	2.035	2.445	2.733
34	1.307	1.691	2.032	2.441	2.728
35	1.306	1.690	2.030	2.438	2.724
36	1.306	1.688	2.028	2.434	2.719
37	1.305	1.687	2.026	2.431	2.715
38	1.304	1.686	2.024	2.429	2.712
39	1.304	1.685	2.023	2.426	2.708
40	1.303	1.684	2.021	2.423	2.704
45	1.301	1.679	2.014	2.412	2.690
50	1.299	1.676	2.009	2.403	2.678
55	1.297	1.673	2.004	2.396	2.668
60	1.296	1.671	2.000	2.390	2.660
65	1.295	1.669	1.997	2.385	2.654
70	1.294	1.667	1.994	2.381	2.648
75	1.293	1.665	1.992	2.377	2.643
80	1.292	1.664	1.990	2.374	2.639
85	1.292	1.663	1.988	2.371	2.635
90	1.291	1.662	1.987	2.368	2.632
95	1.291	1.661	1.985	2.366	2.629
100	1.290	1.660	1.984	2.364	2.626
125	1.288	1.657	1.979	2.357	2.616
150	1.287	1.655	1.976	2.351	2.609
200	1.286	1.653	1.972	2.345	2.601
∞	1.282	1.645	1.960	2.326	2.576

TABLE 8 PERCENTAGE POINTS OF THE F-DISTRIBUTION

The tables give the values of x satisfying

$$P(X \leq x) = p$$

where X is a random variable having the F-distribution with υ_1 degrees of freedom in the numerator and υ_2 degrees of freedom in the denominator.

The table below corresponds to p=0.995 and should be used for one-tail tests at significance level 0.5% or two-tail tests at significance level 1%.

υ_2 \ υ_1	1	2	3	4	5	6	7	8	9	10	11	12	15	20	25	30	40	50	100	∞
1	16211	20000	21615	22500	23056	23437	23715	23925	24091	24224	24334	24426	24630	24836	24960	25044	25148	25211	25337	25465
2	198.5	199.0	199.2	199.2	199.3	199.3	199.4	199.4	199.4	199.4	199.4	199.4	199.4	199.4	199.5	199.5	199.5	199.5	199.5	199.5
3	55.55	49.80	47.47	46.19	45.39	44.84	44.43	44.13	43.88	43.69	43.52	43.39	43.09	42.78	42.59	42.47	42.31	42.21	42.02	41.83
4	31.33	26.28	24.26	23.15	22.46	21.97	21.62	21.35	21.14	20.97	20.82	20.71	20.44	20.17	20.00	19.89	19.75	19.67	19.50	19.32
5	22.78	18.31	16.53	15.56	14.94	14.51	14.20	13.96	13.77	13.62	13.49	13.38	13.15	12.90	12.76	12.66	12.53	12.45	12.30	12.14
6	18.64	14.54	12.92	12.03	11.46	11.07	10.79	10.57	10.39	10.25	10.13	10.03	9.814	9.589	9.451	9.358	9.241	9.170	9.026	8.879
7	16.24	12.40	10.88	10.05	9.522	9.155	8.885	8.678	8.514	8.380	8.270	8.176	7.968	7.754	7.623	7.534	7.422	7.354	7.217	7.076
8	14.69	11.04	9.596	8.805	8.302	7.952	7.694	7.496	7.339	7.211	7.104	7.015	6.814	6.608	6.482	6.396	6.288	6.222	6.088	5.951
9	13.61	10.11	8.717	7.956	7.471	7.134	6.885	6.693	6.541	6.417	6.314	6.227	6.032	5.832	5.708	5.625	5.519	5.454	5.322	5.188
10	12.83	9.427	8.081	7.343	6.872	6.545	6.302	6.116	5.968	5.847	5.746	5.661	5.471	5.274	5.153	5.071	4.966	4.902	4.772	4.639
11	12.23	8.912	7.600	6.881	6.422	6.102	5.865	5.682	5.537	5.418	5.320	5.236	5.049	4.855	4.736	4.654	4.551	4.488	4.359	4.226
12	11.75	8.510	7.226	6.521	6.071	5.757	5.525	5.345	5.202	5.085	4.988	4.906	4.721	4.530	4.412	4.331	4.228	4.165	4.037	3.904
13	11.37	8.186	6.926	6.233	5.791	5.482	5.253	5.076	4.935	4.820	4.724	4.643	4.460	4.270	4.153	4.073	3.970	3.908	3.780	3.647
14	11.06	7.922	6.680	5.998	5.562	5.257	5.031	4.857	4.717	4.603	4.508	4.428	4.247	4.059	3.942	3.862	3.760	3.698	3.569	3.436
15	10.80	7.701	6.476	5.803	5.372	5.071	4.847	4.674	4.536	4.424	4.329	4.250	4.070	3.883	3.766	3.687	3.585	3.523	3.394	3.260
16	10.58	7.514	6.303	5.638	5.212	4.913	4.692	4.521	4.384	4.272	4.179	4.099	3.920	3.734	3.618	3.539	3.437	3.375	3.246	3.112
17	10.38	7.354	6.156	5.497	5.075	4.779	4.559	4.389	4.254	4.142	4.050	3.971	3.793	3.607	3.492	3.412	3.311	3.248	3.119	2.984
18	10.22	7.215	6.028	5.375	4.956	4.663	4.445	4.276	4.141	4.030	3.938	3.860	3.683	3.498	3.382	3.303	3.201	3.139	3.009	2.873
19	10.07	7.093	5.916	5.268	4.853	4.561	4.345	4.177	4.043	3.933	3.841	3.763	3.587	3.402	3.287	3.208	3.106	3.043	2.913	2.776
20	9.944	6.986	5.818	5.174	4.762	4.472	4.257	4.090	3.956	3.847	3.756	3.678	3.502	3.318	3.203	3.123	3.022	2.959	2.828	2.690
25	9.475	6.598	5.462	4.835	4.433	4.150	3.939	3.776	3.645	3.537	3.447	3.370	3.196	3.013	2.898	2.819	2.716	2.652	2.519	2.377
30	9.180	6.355	5.239	4.623	4.228	3.949	3.742	3.580	3.450	3.344	3.255	3.179	3.006	2.823	2.708	2.628	2.524	2.459	2.323	2.176
40	8.828	6.066	4.976	4.374	3.986	3.713	3.509	3.350	3.222	3.117	3.028	2.953	2.781	2.598	2.482	2.401	2.296	2.230	2.088	1.932
50	8.626	5.902	4.826	4.232	3.849	3.579	3.376	3.219	3.092	2.988	2.900	2.825	2.653	2.470	2.353	2.272	2.164	2.097	1.951	1.786
100	8.241	5.589	4.542	3.963	3.589	3.325	3.127	2.972	2.847	2.744	2.657	2.583	2.411	2.227	2.108	2.024	1.912	1.840	1.681	1.485
∞	7.879	5.298	4.279	3.715	3.350	3.091	2.897	2.744	2.621	2.519	2.432	2.358	2.187	2.000	1.877	1.789	1.669	1.590	1.402	1.000

The table below corresponds to p=0.99 and should be used for one-tail tests at significance level 1% or two-tail tests at significance level 2%.

υ_2 \ υ_1	1	2	3	4	5	6	7	8	9	10	11	12	15	20	25	30	40	50	100	∞
1	4052	5000	5403	5625	5764	5859	5928	5981	6022	6056	6083	6106	6158	6209	6240	6261	6287	6303	6334	6366
2	98.50	99.00	99.17	99.25	99.30	99.33	99.36	99.37	99.39	99.40	99.41	99.42	99.43	99.45	99.46	99.47	99.47	99.48	99.49	99.50
3	34.12	30.82	29.46	28.71	28.24	27.91	27.67	27.49	27.35	27.23	27.13	27.05	26.87	26.69	26.58	26.51	26.41	26.35	26.24	26.13
4	21.20	18.00	16.69	15.98	15.52	15.21	14.98	14.80	14.66	14.55	14.45	14.37	14.20	14.02	13.91	13.84	13.75	13.69	13.58	13.46
5	16.26	13.27	12.06	11.39	10.97	10.67	10.46	10.29	10.16	10.05	9.963	9.888	9.722	9.553	9.449	9.379	9.291	9.238	9.130	9.020
6	13.75	10.93	9.780	9.148	8.746	8.466	8.260	8.102	7.976	7.874	7.790	7.718	7.559	7.396	7.296	7.229	7.143	7.091	6.987	6.880
7	12.25	9.547	8.451	7.847	7.460	7.191	6.993	6.840	6.719	6.620	6.538	6.469	6.314	6.155	6.058	5.992	5.908	5.858	5.755	5.650
8	11.26	8.649	7.591	7.006	6.632	6.371	6.178	6.029	5.911	5.814	5.734	5.667	5.515	5.359	5.263	5.198	5.116	5.065	4.963	4.860
9	10.56	8.022	6.992	6.422	6.057	5.802	5.613	5.467	5.351	5.257	5.178	5.111	4.962	4.808	4.713	4.649	4.567	4.517	4.415	4.311
10	10.04	7.559	6.552	5.994	5.636	5.386	5.200	5.057	4.942	4.849	4.772	4.706	4.558	4.405	4.311	4.247	4.165	4.115	4.014	3.909
11	9.646	7.206	6.217	5.668	5.316	5.069	4.886	4.744	4.632	4.539	4.462	4.397	4.251	4.099	4.005	3.941	3.860	3.810	3.708	3.602
12	9.330	6.927	5.953	5.412	5.064	4.821	4.640	4.499	4.388	4.296	4.220	4.155	4.010	3.858	3.765	3.701	3.619	3.569	3.467	3.361
13	9.074	6.701	5.739	5.205	4.862	4.620	4.441	4.302	4.191	4.100	4.025	3.960	3.815	3.665	3.571	3.507	3.425	3.375	3.272	3.165
14	8.862	6.515	5.564	5.035	4.695	4.456	4.278	4.140	4.030	3.939	3.864	3.800	3.656	3.505	3.412	3.348	3.266	3.215	3.112	3.004
15	8.683	6.359	5.417	4.893	4.556	4.318	4.142	4.004	3.895	3.805	3.730	3.666	3.522	3.372	3.278	3.214	3.132	3.081	2.977	2.868
16	8.531	6.226	5.292	4.773	4.437	4.202	4.026	3.890	3.780	3.691	3.616	3.553	3.409	3.259	3.165	3.101	3.018	2.967	2.863	2.753
17	8.400	6.112	5.185	4.669	4.336	4.102	3.927	3.791	3.682	3.593	3.519	3.455	3.312	3.162	3.068	3.003	2.920	2.869	2.764	2.653
18	8.285	6.013	5.092	4.579	4.248	4.015	3.841	3.705	3.597	3.508	3.434	3.371	3.227	3.077	2.983	2.919	2.835	2.784	2.678	2.566
19	8.185	5.926	5.010	4.500	4.171	3.939	3.765	3.631	3.523	3.434	3.360	3.297	3.153	3.003	2.909	2.844	2.761	2.709	2.602	2.489
20	8.096	5.849	4.938	4.431	4.103	3.871	3.699	3.564	3.457	3.368	3.294	3.231	3.088	2.938	2.843	2.778	2.695	2.643	2.535	2.421
25	7.770	5.568	4.675	4.177	3.855	3.627	3.457	3.324	3.217	3.129	3.056	2.993	2.850	2.699	2.604	2.538	2.453	2.400	2.289	2.169
30	7.562	5.390	4.510	4.018	3.699	3.473	3.304	3.173	3.067	2.979	2.906	2.843	2.700	2.549	2.453	2.386	2.299	2.245	2.131	2.006
40	7.314	5.179	4.313	3.828	3.514	3.291	3.124	2.993	2.888	2.801	2.727	2.665	2.522	2.369	2.271	2.203	2.114	2.058	1.938	1.805
50	7.171	5.057	4.199	3.720	3.408	3.186	3.020	2.890	2.785	2.698	2.625	2.562	2.419	2.265	2.167	2.098	2.007	1.949	1.825	1.683
100	6.895	4.824	3.984	3.513	3.206	2.988	2.823	2.694	2.590	2.503	2.430	2.368	2.223	2.067	1.965	1.893	1.797	1.735	1.598	1.427
∞	6.635	4.605	3.782	3.319	3.017	2.802	2.639	2.511	2.407	2.321	2.248	2.185	2.039	1.878	1.773	1.696	1.592	1.523	1.358	1.000

PERCENTAGE POINTS OF THE F-DISTRIBUTION

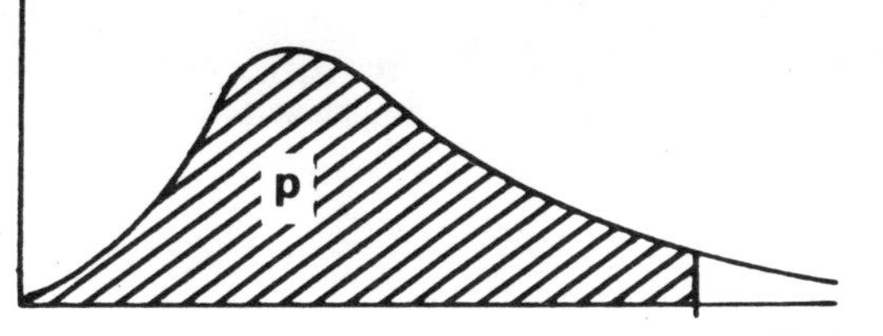

$$x = F_p(\upsilon_1, \upsilon_2)$$

The relationship

$$F_p(\upsilon_1, \upsilon_2) = 1/F_{1-p}(\upsilon_2, \upsilon_1)$$

can be used to find the percentage points in the lower tail.

The table below corresponds to $p = 0.975$ and should be used for one-tail tests at significance level 2.5% or two-tail tests at significance level 5%.

υ_2 \ υ_1	1	2	3	4	5	6	7	8	9	10	11	12	15	20	25	30	40	50	100	∞
1	647.8	799.5	864.2	899.6	921.8	937.1	948.2	956.7	963.3	968.6	973.0	976.7	984.9	993.1	998.1	1001	1006	1008	1013	1018
2	38.51	39.00	39.17	39.25	39.30	39.33	39.36	39.37	39.39	39.40	39.41	39.42	39.43	39.45	39.46	39.47	39.47	39.48	39.49	39.50
3	17.44	16.04	15.44	15.10	14.89	14.74	14.62	14.54	14.47	14.42	14.37	14.34	14.25	14.17	14.12	14.08	14.04	14.01	13.96	13.90
4	12.22	10.65	9.979	9.605	9.364	9.197	9.074	8.980	8.905	8.844	8.794	8.751	8.657	8.560	8.501	8.461	8.411	8.381	8.319	8.257
5	10.01	8.434	7.764	7.388	7.146	6.978	6.853	6.757	6.681	6.619	6.568	6.525	6.428	6.329	6.268	6.227	6.175	6.144	6.080	6.015
6	8.813	7.260	6.599	6.227	5.988	5.820	5.695	5.600	5.523	5.461	5.410	5.366	5.269	5.168	5.107	5.065	5.012	4.980	4.915	4.849
7	8.073	6.542	5.890	5.523	5.285	5.119	4.995	4.899	4.823	4.761	4.709	4.666	4.568	4.467	4.405	4.362	4.309	4.276	4.210	4.142
8	7.571	6.059	5.416	5.053	4.817	4.652	4.529	4.433	4.357	4.295	4.243	4.200	4.101	3.999	3.937	3.894	3.840	3.807	3.739	3.670
9	7.209	5.715	5.078	4.718	4.484	4.320	4.197	4.102	4.026	3.964	3.912	3.868	3.769	3.667	3.604	3.560	3.505	3.472	3.403	3.333
10	6.937	5.456	4.826	4.468	4.236	4.072	3.950	3.855	3.779	3.717	3.665	3.621	3.522	3.419	3.355	3.311	3.255	3.221	3.152	3.080
11	6.724	5.256	4.630	4.275	4.044	3.881	3.759	3.664	3.588	3.526	3.474	3.430	3.330	3.226	3.162	3.118	3.061	3.027	2.956	2.883
12	6.554	5.096	4.474	4.121	3.891	3.728	3.607	3.512	3.436	3.374	3.321	3.277	3.177	3.073	3.008	2.963	2.906	2.871	2.800	2.725
13	6.414	4.965	4.347	3.996	3.767	3.604	3.483	3.388	3.312	3.250	3.197	3.153	3.053	2.948	2.882	2.837	2.780	2.744	2.671	2.595
14	6.298	4.857	4.242	3.892	3.663	3.501	3.380	3.285	3.209	3.147	3.095	3.050	2.949	2.844	2.778	2.732	2.674	2.638	2.565	2.487
15	6.200	4.765	4.153	3.804	3.576	3.415	3.293	3.199	3.123	3.060	3.008	2.963	2.862	2.756	2.689	2.644	2.585	2.549	2.474	2.395
16	6.115	4.687	4.077	3.729	3.502	3.341	3.219	3.125	3.049	2.986	2.934	2.889	2.788	2.681	2.614	2.568	2.509	2.472	2.396	2.316
17	6.042	4.619	4.011	3.665	3.438	3.277	3.156	3.061	2.985	2.922	2.870	2.825	2.723	2.616	2.548	2.502	2.442	2.405	2.329	2.247
18	5.978	4.560	3.954	3.608	3.382	3.221	3.100	3.005	2.929	2.866	2.814	2.769	2.667	2.559	2.491	2.445	2.384	2.347	2.269	2.187
19	5.922	4.508	3.903	3.559	3.333	3.172	3.051	2.956	2.880	2.817	2.765	2.720	2.617	2.509	2.441	2.394	2.333	2.295	2.217	2.133
20	5.871	4.461	3.859	3.515	3.289	3.128	3.007	2.913	2.837	2.774	2.721	2.676	2.573	2.464	2.396	2.349	2.287	2.249	2.170	2.085
25	5.686	4.291	3.694	3.353	3.129	2.969	2.848	2.753	2.677	2.613	2.560	2.515	2.411	2.300	2.230	2.182	2.118	2.079	1.996	1.906
30	5.568	4.182	3.589	3.250	3.026	2.867	2.746	2.651	2.575	2.511	2.458	2.412	2.307	2.195	2.124	2.074	2.009	1.968	1.882	1.787
40	5.424	4.051	3.463	3.126	2.904	2.744	2.624	2.529	2.452	2.388	2.334	2.288	2.182	2.068	1.994	1.943	1.875	1.832	1.741	1.637
50	5.340	3.975	3.390	3.054	2.833	2.674	2.553	2.458	2.381	2.317	2.263	2.216	2.109	1.993	1.919	1.866	1.796	1.752	1.656	1.545
100	5.179	3.828	3.250	2.917	2.696	2.537	2.417	2.321	2.244	2.179	2.124	2.077	1.968	1.849	1.770	1.715	1.640	1.592	1.483	1.347
∞	5.024	3.689	3.116	2.786	2.567	2.408	2.288	2.192	2.114	2.048	1.993	1.945	1.833	1.708	1.626	1.566	1.484	1.428	1.296	1.000

The table below corresponds to $p = 0.95$ and should be used for one-tail tests at significance level 5% or two-tail tests at significance level 10%.

υ_2 \ υ_1	1	2	3	4	5	6	7	8	9	10	11	12	15	20	25	30	40	50	100	∞
1	161.4	199.5	215.7	224.6	230.2	234.0	236.8	238.9	240.5	241.9	243.0	243.9	246.0	248.0	249.3	250.1	251.1	251.8	253.0	245.3
2	18.51	19.00	19.16	19.25	19.30	19.33	19.35	19.37	19.39	19.40	19.41	19.41	19.43	19.45	19.46	19.46	19.47	19.48	19.49	19.50
3	10.13	9.552	9.277	9.117	9.013	8.941	8.887	8.845	8.812	8.786	8.763	8.745	8.703	8.660	8.634	8.617	8.594	8.581	8.554	8.526
4	7.709	6.944	6.591	6.388	6.256	6.163	6.094	6.041	5.999	5.964	5.936	5.912	5.858	5.803	5.769	5.746	5.717	5.699	5.664	5.628
5	6.608	5.786	5.409	5.192	5.050	4.950	4.876	4.818	4.772	4.735	4.704	4.678	4.619	4.558	4.521	4.496	4.464	4.444	4.405	4.365
6	5.987	5.143	4.757	4.534	4.387	4.284	4.207	4.147	4.099	4.060	4.027	4.000	3.938	3.874	3.835	3.808	3.774	3.754	3.712	3.669
7	5.591	4.737	4.347	4.120	3.972	3.866	3.787	3.726	3.677	3.637	3.603	3.575	3.511	3.445	3.404	3.376	3.340	3.319	3.275	3.230
8	5.318	4.459	4.066	3.838	3.687	3.581	3.500	3.438	3.388	3.347	3.313	3.284	3.218	3.150	3.108	3.079	3.043	3.020	2.975	2.928
9	5.117	4.256	3.863	3.633	3.482	3.374	3.293	3.230	3.179	3.137	3.102	3.073	3.006	2.936	2.893	2.864	2.826	2.803	2.756	2.707
10	4.965	4.103	3.708	3.478	3.326	3.217	3.135	3.072	3.020	2.978	2.943	2.913	2.845	2.774	2.730	2.700	2.661	2.637	2.588	2.538
11	4.844	3.982	3.587	3.357	3.204	3.095	3.012	2.948	2.896	2.854	2.818	2.788	2.719	2.646	2.601	2.570	2.531	2.507	2.457	2.404
12	4.747	3.885	3.490	3.259	3.106	2.996	2.913	2.849	2.796	2.753	2.717	2.687	2.617	2.544	2.498	2.466	2.426	2.401	2.350	2.296
13	4.667	3.806	3.411	3.179	3.025	2.915	2.832	2.767	2.714	2.671	2.635	2.604	2.533	2.459	2.412	2.380	2.339	2.314	2.261	2.206
14	4.600	3.739	3.344	3.112	2.958	2.848	2.764	2.699	2.646	2.602	2.565	2.534	2.463	2.388	2.341	2.308	2.266	2.241	2.187	2.131
15	4.543	3.682	3.287	3.056	2.901	2.790	2.707	2.641	2.588	2.544	2.507	2.475	2.403	2.328	2.280	2.247	2.204	2.178	2.123	2.066
16	4.494	3.634	3.239	3.007	2.852	2.741	2.657	2.591	2.538	2.494	2.456	2.425	2.352	2.276	2.227	2.194	2.151	2.124	2.068	2.010
17	4.451	3.592	3.197	2.965	2.810	2.699	2.614	2.548	2.494	2.450	2.413	2.381	2.308	2.230	2.181	2.148	2.104	2.077	2.020	1.960
18	4.414	3.555	3.160	2.928	2.773	2.661	2.577	2.510	2.456	2.412	2.374	2.342	2.269	2.191	2.141	2.107	2.063	2.035	1.978	1.917
19	4.381	3.522	3.127	2.895	2.740	2.628	2.544	2.477	2.423	2.378	2.340	2.308	2.234	2.155	2.106	2.071	2.026	1.999	1.940	1.878
20	4.351	3.493	3.098	2.866	2.711	2.599	2.514	2.447	2.393	2.348	2.310	2.278	2.203	2.124	2.074	2.039	1.994	1.966	1.907	1.843
25	4.242	3.385	2.991	2.759	2.603	2.490	2.405	2.337	2.282	2.236	2.198	2.165	2.089	2.007	1.955	1.919	1.872	1.842	1.779	1.711
30	4.171	3.316	2.922	2.690	2.534	2.421	2.334	2.266	2.211	2.165	2.126	2.092	2.015	1.932	1.878	1.841	1.792	1.761	1.695	1.622
40	4.085	3.232	2.839	2.606	2.449	2.336	2.249	2.180	2.124	2.077	2.038	2.003	1.924	1.839	1.783	1.744	1.693	1.660	1.589	1.509
50	4.034	3.183	2.790	2.557	2.400	2.286	2.199	2.130	2.073	2.026	1.986	1.952	1.871	1.784	1.727	1.687	1.634	1.599	1.525	1.438
100	3.936	3.087	2.696	2.463	2.305	2.191	2.103	2.032	1.975	1.927	1.886	1.850	1.768	1.676	1.616	1.573	1.515	1.477	1.392	1.283
∞	3.841	2.996	2.605	2.372	2.214	2.099	2.010	1.938	1.880	1.831	1.789	1.752	1.666	1.571	1.506	1.459	1.394	1.350	1.243	1.000

TABLE 9 CRITICAL VALUES OF THE PRODUCT MOMENT CORRELATION COEFFICIENT

The table gives the critical values, for different significance levels, of the sample product moment correlation coefficient r based on n independent pairs of observations from a bivariate normal distribution with correlation coefficient $\rho = 0$.

One tail Two tail n	10% 20%	5% 10%	2.5% 5%	1% 2%	0.5% 1%
4	0.8000	0.9000	0.9500	0.9800	0.9900
5	0.6870	0.8054	0.8783	0.9343	0.9587
6	0.6084	0.7293	0.8114	0.8822	0.9172
7	0.5509	0.6694	0.7545	0.8329	0.8745
8	0.5067	0.6215	0.7067	0.7887	0.8343
9	0.4716	0.5822	0.6664	0.7498	0.7977
10	0.4428	0.5494	0.6319	0.7155	0.7646
11	0.4187	0.5214	0.6021	0.6851	0.7348
12	0.3981	0.4973	0.5760	0.6581	0.7079
13	0.3802	0.4762	0.5529	0.6339	0.6835
14	0.3646	0.4575	0.5324	0.6120	0.6614
15	0.3507	0.4409	0.5140	0.5923	0.6411
16	0.3383	0.4259	0.4973	0.5742	0.6226
17	0.3271	0.4124	0.4821	0.5577	0.6055
18	0.3170	0.4000	0.4683	0.5425	0.5897
19	0.3077	0.3887	0.4555	0.5285	0.5751
20	0.2992	0.3783	0.4438	0.5155	0.5614
21	0.2914	0.3687	0.4329	0.5034	0.5487
22	0.2841	0.3598	0.4227	0.4921	0.5368
23	0.2774	0.3515	0.4132	0.4815	0.5256
24	0.2711	0.3438	0.4044	0.4716	0.5151
25	0.2653	0.3365	0.3961	0.4622	0.5052
26	0.2598	0.3297	0.3882	0.4534	0.4958
27	0.2546	0.3233	0.3809	0.4451	0.4869
28	0.2497	0.3172	0.3739	0.4372	0.4785
29	0.2451	0.3115	0.3673	0.4297	0.4705
30	0.2407	0.3061	0.3610	0.4226	0.4629
31	0.2366	0.3009	0.3550	0.4158	0.4556
32	0.2327	0.2960	0.3494	0.4093	0.4487
33	0.2289	0.2913	0.3440	0.4032	0.4421
34	0.2254	0.2869	0.3388	0.3972	0.4357
35	0.2220	0.2826	0.3338	0.3916	0.4296
36	0.2187	0.2785	0.3291	0.3862	0.4238
37	0.2156	0.2746	0.3246	0.3810	0.4182
38	0.2126	0.2709	0.3202	0.3760	0.4128
39	0.2097	0.2673	0.3160	0.3712	0.4076
40	0.2070	0.2638	0.3120	0.3665	0.4026
41	0.2043	0.2605	0.3081	0.3621	0.3978
42	0.2018	0.2573	0.3044	0.3578	0.3932
43	0.1993	0.2542	0.3008	0.3536	0.3887
44	0.1970	0.2512	0.2973	0.3496	0.3843
45	0.1947	0.2483	0.2940	0.3457	0.3801
46	0.1925	0.2455	0.2907	0.3420	0.3761
47	0.1903	0.2429	0.2876	0.3384	0.3721
48	0.1883	0.2403	0.2845	0.3348	0.3683
49	0.1863	0.2377	0.2816	0.3314	0.3646
50	0.1843	0.2353	0.2787	0.3281	0.3610
60	0.1678	0.2144	0.2542	0.2997	0.3301
70	0.1550	0.1982	0.2352	0.2776	0.3060
80	0.1448	0.1852	0.2199	0.2597	0.2864
90	0.1364	0.1745	0.2072	0.2449	0.2702
100	0.1292	0.1654	0.1966	0.2324	0.2565

TABLE 10 CRITICAL VALUES OF THE SPEARMAN RANK CORRELATION COEFFICIENT

The table gives the critical values, for different significance levels, of the Spearman rank correlation coefficient r_s for various sample sizes n. It should be noted that, since r_s is discrete, exact significance levels cannot in general be achieved. The critical values given are those whose significance levels are nearest to the stated values.

One tail Two tail n	10% 20%	5% 10%	2.5% 5%	1% 2%	0.5% 1%
4	1.0000	1.0000	1.0000	1.0000	1.0000
5	0.7000	0.9000	0.9000	1.0000	1.0000
6	0.6571	0.7714	0.8286	0.9429	0.9429
7	0.5714	0.6786	0.7857	0.8571	0.8929
8	0.5476	0.6429	0.7381	0.8095	0.8571
9	0.4833	0.6000	0.6833	0.7667	0.8167
10	0.4424	0.5636	0.6485	0.7333	0.7818
11	0.4182	0.5273	0.6091	0.7000	0.7545
12	0.3986	0.5035	0.5874	0.6713	0.7273
13	0.3791	0.4780	0.5604	0.6484	0.6978
14	0.3670	0.4593	0.5385	0.6220	0.6747
15	0.3500	0.4429	0.5179	0.6000	0.6536
16	0.3382	0.4265	0.5029	0.5824	0.6324
17	0.3271	0.4124	0.4821	0.5577	0.6055
18	0.3170	0.4000	0.4683	0.5425	0.5897
19	0.3077	0.3887	0.4555	0.5285	0.5751
20	0.2992	0.3783	0.4438	0.5155	0.5614
21	0.2914	0.3687	0.4329	0.5034	0.5487
22	0.2841	0.3598	0.4227	0.4921	0.5368
23	0.2774	0.3515	0.4132	0.4815	0.5256
24	0.2711	0.3438	0.4044	0.4716	0.5151
25	0.2653	0.3365	0.3961	0.4622	0.5052
26	0.2598	0.3297	0.3882	0.4534	0.4958
27	0.2546	0.3233	0.3809	0.4451	0.4869
28	0.2497	0.3172	0.3739	0.4372	0.4785
29	0.2451	0.3115	0.3673	0.4297	0.4705
30	0.2407	0.3061	0.3610	0.4226	0.4629
31	0.2366	0.3009	0.3550	0.4158	0.4556
32	0.2327	0.2960	0.3494	0.4093	0.4487
33	0.2289	0.2913	0.3440	0.4032	0.4421
34	0.2254	0.2869	0.3388	0.3972	0.4357
35	0.2220	0.2826	0.3338	0.3916	0.4296
36	0.2187	0.2785	0.3291	0.3862	0.4238
37	0.2156	0.2746	0.3246	0.3810	0.4182
38	0.2126	0.2709	0.3202	0.3760	0.4128
39	0.2097	0.2673	0.3160	0.3712	0.4076
40	0.2070	0.2638	0.3120	0.3665	0.4026
41	0.2043	0.2605	0.3081	0.3621	0.3978
42	0.2018	0.2573	0.3044	0.3578	0.3932
43	0.1993	0.2542	0.3008	0.3536	0.3887
44	0.1970	0.2512	0.2973	0.3496	0.3843
45	0.1947	0.2483	0.2940	0.3457	0.3801
46	0.1925	0.2455	0.2907	0.3420	0.3761
47	0.1903	0.2429	0.2876	0.3384	0.3721
48	0.1883	0.2403	0.2845	0.3348	0.3683
49	0.1863	0.2377	0.2816	0.3314	0.3646
50	0.1843	0.2353	0.2787	0.3281	0.3610
60	0.1678	0.2144	0.2542	0.2997	0.3301
70	0.1550	0.1982	0.2352	0.2776	0.3060
80	0.1448	0.1852	0.2199	0.2597	0.2864
90	0.1364	0.1745	0.2072	0.2449	0.2702
100	0.1292	0.1654	0.1966	0.2324	0.2565

TABLE 11 THE FISHER z-TRANSFORMATION

The table gives the values of the function $z(r) = \tanh^{-1} r$. For $r < 0$, the relationship $z(r) = -z(-r)$ may be used.

r	.00	.01	.02	.03	.04	.05	.06	.07	.08	.09
0.00	0.0000	0.0100	0.0200	0.0300	0.0400	0.0500	0.0601	0.0701	0.0802	0.0902
0.10	0.1003	0.1104	0.1206	0.1307	0.1409	0.1511	0.1614	0.1717	0.1820	0.1923
0.20	0.2027	0.2132	0.2237	0.2342	0.2448	0.2554	0.2661	0.2769	0.2877	0.2986
0.30	0.3095	0.3205	0.3316	0.3428	0.3541	0.3654	0.3769	0.3884	0.4001	0.4118
0.40	0.4236	0.4356	0.4477	0.4599	0.4722	0.4847	0.4973	0.5101	0.5230	0.5361
0.50	0.5493	0.5627	0.5763	0.5901	0.6042	0.6184	0.6328	0.6475	0.6625	0.6777
0.60	0.6931	0.7089	0.7250	0.7414	0.7582	0.7753	0.7928	0.8107	0.8291	0.8480
0.70	0.8673	0.8872	0.9076	0.9287	0.9505	0.9730	0.9962	1.0203	1.0454	1.0714
0.80	1.0986	1.1270	1.1568	1.1881	1.2212	1.2562	1.2933	1.3331	1.3758	1.4219

r	.000	.001	.002	.003	.004	.005	.006	.007	.008	.009
0.900	1.4722	1.4775	1.4828	1.4882	1.4937	1.4992	1.5047	1.5103	1.5160	1.5217
0.910	1.5275	1.5334	1.5393	1.5453	1.5513	1.5574	1.5636	1.5698	1.5762	1.5826
0.920	1.5890	1.5956	1.6022	1.6089	1.6157	1.6226	1.6296	1.6366	1.6438	1.6510
0.930	1.6584	1.6658	1.6734	1.6811	1.6888	1.6967	1.7047	1.7129	1.7211	1.7295
0.940	1.7380	1.7467	1.7555	1.7645	1.7736	1.7828	1.7923	1.8019	1.8117	1.8216
0.950	1.8318	1.8421	1.8527	1.8635	1.8745	1.8857	1.8972	1.9090	1.9210	1.9333
0.960	1.9459	1.9588	1.9721	1.9857	1.9996	2.0139	2.0287	2.0439	2.0595	2.0756
0.970	2.0923	2.1095	2.1273	2.1457	2.1649	2.1847	2.2054	2.2269	2.2494	2.2729
0.980	2.2976	2.3235	2.3507	2.3796	2.4101	2.4427	2.4774	2.5147	2.5550	2.5987
0.990	2.6467	2.6996	2.7587	2.8257	2.9031	2.9945	3.1063	3.2504	3.4534	3.8002

TABLE 12 THE INVERSE FISHER z-TRANSFORMATION

The table gives the values of the function $r(z) = \tanh z$. For $z < 0$, the relationship $r(z) = -r(-z)$ may be used.

z	.00	.01	.02	.03	.04	.05	.06	.07	.08	.09
0.00	0.0000	0.0100	0.0200	0.0300	0.0400	0.0500	0.0599	0.0699	0.0798	0.0898
0.10	0.0997	0.1096	0.1194	0.1293	0.1391	0.1489	0.1586	0.1684	0.1781	0.1877
0.20	0.1974	0.2070	0.2165	0.2260	0.2355	0.2449	0.2543	0.2636	0.2729	0.2821
0.30	0.2913	0.3004	0.3095	0.3185	0.3275	0.3364	0.3452	0.3540	0.3627	0.3714
0.40	0.3799	0.3885	0.3969	0.4053	0.4136	0.4219	0.4301	0.4382	0.4462	0.4542
0.50	0.4621	0.4699	0.4777	0.4854	0.4930	0.5005	0.5080	0.5154	0.5227	0.5299
0.60	0.5370	0.5441	0.5511	0.5581	0.5649	0.5717	0.5784	0.5850	0.5915	0.5980
0.70	0.6044	0.6107	0.6169	0.6231	0.6291	0.6351	0.6411	0.6469	0.6527	0.6584
0.80	0.6640	0.6696	0.6751	0.6805	0.6858	0.6911	0.6963	0.7014	0.7064	0.7114
0.90	0.7163	0.7211	0.7259	0.7306	0.7352	0.7398	0.7443	0.7487	0.7531	0.7574
1.00	0.7616	0.7658	0.7699	0.7739	0.7779	0.7818	0.7857	0.7895	0.7932	0.7969
1.10	0.8005	0.8041	0.8076	0.8110	0.8144	0.8178	0.8210	0.8243	0.8275	0.8306
1.20	0.8337	0.8367	0.8397	0.8426	0.8455	0.8483	0.8511	0.8538	0.8565	0.8591
1.30	0.8617	0.8643	0.8668	0.8692	0.8717	0.8741	0.8764	0.8787	0.8810	0.8832
1.40	0.8854	0.8875	0.8896	0.8917	0.8937	0.8957	0.8977	0.8996	0.9015	0.9033
1.50	0.9051	0.9069	0.9087	0.9104	0.9121	0.9138	0.9154	0.9170	0.9186	0.9201
1.60	0.9217	0.9232	0.9246	0.9261	0.9275	0.9289	0.9302	0.9316	0.9329	0.9341
1.70	0.9354	0.9366	0.9379	0.9391	0.9402	0.9414	0.9425	0.9436	0.9447	0.9458
1.80	0.9468	0.9478	0.9488	0.9498	0.9508	0.9517	0.9527	0.9536	0.9545	0.9554
1.90	0.9562	0.9571	0.9579	0.9587	0.9595	0.9603	0.9611	0.9618	0.9626	0.9633
2.00	0.9640	0.9647	0.9654	0.9661	0.9667	0.9674	0.9680	0.9687	0.9693	0.9699
2.10	0.9705	0.9710	0.9716	0.9721	0.9727	0.9732	0.9737	0.9743	0.9748	0.9753
2.20	0.9757	0.9762	0.9767	0.9771	0.9776	0.9780	0.9785	0.9789	0.9793	0.9797
2.30	0.9801	0.9805	0.9809	0.9812	0.9816	0.9820	0.9823	0.9827	0.9830	0.9833
2.40	0.9837	0.9840	0.9843	0.9846	0.9849	0.9852	0.9855	0.9858	0.9861	0.9863
2.50	0.9866	0.9869	0.9871	0.9874	0.9876	0.9879	0.9881	0.9884	0.9886	0.9888
2.60	0.9890	0.9892	0.9895	0.9897	0.9899	0.9901	0.9903	0.9905	0.9906	0.9908
2.70	0.9910	0.9912	0.9914	0.9915	0.9917	0.9919	0.9920	0.9922	0.9923	0.9925
2.80	0.9926	0.9928	0.9929	0.9931	0.9932	0.9933	0.9935	0.9936	0.9937	0.9938
2.90	0.9940	0.9941	0.9942	0.9943	0.9944	0.9945	0.9946	0.9947	0.9949	0.9950
3.00	0.9951	0.9952	0.9952	0.9953	0.9954	0.9955	0.9956	0.9957	0.9958	0.9959
3.10	0.9959	0.9960	0.9961	0.9962	0.9963	0.9963	0.9964	0.9965	0.9965	0.9966
3.20	0.9967	0.9967	0.9968	0.9969	0.9969	0.9970	0.9971	0.9971	0.9972	0.9972
3.30	0.9973	0.9973	0.9974	0.9974	0.9975	0.9975	0.9976	0.9976	0.9977	0.9977
3.40	0.9978	0.9978	0.9979	0.9979	0.9979	0.9980	0.9980	0.9981	0.9981	0.9981
3.50	0.9982	0.9982	0.9982	0.9983	0.9983	0.9984	0.9984	0.9984	0.9984	0.9985
3.60	0.9985	0.9985	0.9986	0.9986	0.9986	0.9986	0.9987	0.9987	0.9987	0.9988
3.70	0.9988	0.9988	0.9988	0.9988	0.9989	0.9989	0.9989	0.9989	0.9990	0.9990
3.80	0.9990	0.9990	0.9990	0.9991	0.9991	0.9991	0.9991	0.9991	0.9991	0.9992
3.90	0.9992	0.9992	0.9992	0.9992	0.9992	0.9993	0.9993	0.9993	0.9993	0.9993

TABLE 13 CRITICAL VALUES OF THE WILCOXON SIGNED RANK STATISTIC

The table gives the upper tail critical values w_c of the statistic

$$W = \sum_{i=1}^{n} U_i R_i$$

where R_i denotes the rank of the magnitude of the ith. observation in a sample of size n and $U_i = 1$ or 0 according as to whether this observation is positive or negative. The lower tail critical values are given by $\frac{1}{2}n(n+1) - w_c$ .Since W is discrete, exact significance levels cannot in general be achieved. The critical values given are those whose significance levels are nearest to those stated.

One tail	10%	5%	2.5%	1%	0.5%
Two tail	20%	10%	5%	2%	1%
n					
3	6				
4	9	10			
5	13	14	15		
6	17	19	20	21	
7	22	24	26	28	28
8	28	30	32	34	36
9	34	37	39	42	43
10	41	44	47	50	52
11	48	52	55	59	61
12	56	61	64	68	71
13	65	70	74	78	81
14	74	79	84	89	92
15	83	90	95	100	104
16	94	100	106	112	117
17	104	112	118	125	130
18	116	124	131	138	143
19	128	136	144	152	158
20	140	150	158	167	173
21	153	163	172	182	188
22	167	178	187	197	204
23	181	193	203	214	221
24	196	208	219	231	239
25	211	224	235	248	257
26	227	241	253	266	275
27	243	258	271	285	294
28	260	276	289	304	314
29	278	294	308	324	335
30	296	313	328	345	356
32	333	353	369	387	400
34	373	394	412	433	446
36	416	438	458	480	495
38	460	485	506	530	546
40	506	533	556	582	599
45	632	664	691	722	743
50	771	809	841	877	902

TABLE 14 CRITICAL VALUES OF THE MANN-WHITNEY STATISTIC

The table gives the upper tail critical values u_c of the statistic

$$U = \sum_{i=1}^{m} \sum_{j=1}^{n} Z_{ij}$$

where $Z_{ij} = 1$ if $X_i < Y_j$ and $Z_{ij} = 0$ if $X_i > Y_j$ given the independent samples $X_1, X_2, \ldots. X_m$ and $Y_1, Y_2, \ldots. Y_n$. The lower tail critical values are given by $mn - u_c$.

One tail 0.5% Two tail 1%

n \ m	1	2	3	4	5	6	7	8	9	10	11	12	13	14	15	16	17	18	19	20	21	22	23	24	25	26	27	28	29	30
1																														
2													26	28	30	32	34	36	38	40	42	44	45	47	49	51	53	55	57	59
3							21	24	27	30	32	35	38	40	43	46	48	51	54	57	59	62	65	67	70	73	75	78	81	83
4					20	24	27	31	34	38	41	45	48	51	55	58	62	65	69	72	75	79	82	86	89	93	96	99	103	106
5				20	25	29	33	37	41	46	50	54	58	62	66	70	74	79	83	87	91	95	99	103	107	112	116	120	124	128
6				24	29	34	39	44	48	53	58	63	68	73	77	82	87	92	97	101	106	111	116	120	125	130	135	140	144	149
7			21	27	33	39	44	50	55	61	66	72	77	83	88	94	99	105	110	116	121	127	132	137	143	148	154	159	165	170
8			24	31	37	44	50	56	62	69	75	81	87	93	99	105	111	118	124	130	136	142	148	154	160	166	172	178	185	191
9			27	34	41	48	55	62	69	76	83	90	97	103	110	117	124	130	137	144	151	157	164	171	177	184	191	198	204	211
10			30	38	46	53	61	69	76	84	91	99	106	113	121	128	136	143	150	158	165	172	180	187	195	202	209	217	224	231
11			32	41	50	58	66	75	83	91	99	107	115	123	131	140	148	156	164	172	180	188	196	204	212	220	227	235	243	251
12			35	45	54	63	72	81	90	99	107	116	125	133	142	151	159	168	177	185	194	203	211	220	228	237	246	254	263	271
13		26	38	48	58	68	77	87	97	106	115	125	134	143	153	162	171	181	190	199	208	218	227	236	245	254	264	273	282	291
14		28	40	51	62	73	83	93	103	113	123	133	143	153	163	173	183	193	203	213	223	233	242	252	262	272	282	292	301	311
15		30	43	55	66	77	88	99	110	121	131	142	153	163	174	184	195	206	216	226	237	247	258	268	279	289	300	310	320	331
16		32	46	58	70	82	94	105	117	128	140	151	162	173	184	196	207	218	229	240	251	262	273	284	295	307	318	329	340	351
17		34	48	62	74	87	99	111	124	136	148	159	171	183	195	207	219	230	242	254	265	277	289	300	312	324	335	347	359	370
18		36	51	65	79	92	105	118	130	143	156	168	181	193	206	218	230	243	255	267	280	292	304	316	329	341	353	365	378	390
19		38	54	69	83	97	110	124	137	150	164	177	190	203	216	229	242	255	268	281	294	307	320	332	345	358	371	384	397	409
20		40	57	72	87	101	116	130	144	158	172	185	199	213	226	240	254	267	281	294	308	321	335	348	362	375	389	402	416	429
21		42	59	75	91	106	121	136	151	165	180	194	208	223	237	251	265	280	294	308	322	336	350	364	378	392	406	420	435	449
22		44	62	79	95	111	127	142	157	172	188	203	218	233	247	262	277	292	307	321	336	351	366	380	395	410	424	439	453	468
23		45	65	82	99	116	132	148	164	180	196	211	227	242	258	273	289	304	320	335	350	366	381	396	411	427	442	457	472	487
24		47	67	86	103	120	137	154	171	187	204	220	236	252	268	284	300	316	332	348	364	380	396	412	428	444	459	475	491	507
25		49	70	89	107	125	143	160	177	195	212	228	245	262	279	295	312	329	345	362	378	395	411	428	444	461	477	494	510	526
26		51	73	93	112	130	148	166	184	202	220	237	254	272	289	307	324	341	358	375	392	410	427	444	461	478	495	512	529	546
27		53	75	96	116	135	154	172	191	209	227	246	264	282	300	318	335	353	371	389	406	424	442	459	477	495	512	530	548	565
28		55	78	99	120	140	159	178	198	217	235	254	273	292	310	329	347	365	384	402	420	439	457	475	494	512	530	548	566	584
29		57	81	103	124	144	165	185	204	224	243	263	282	301	320	340	359	378	397	416	435	453	472	491	510	529	548	566	585	604
30		59	83	106	128	149	170	191	211	231	251	271	291	311	331	351	370	390	409	429	449	468	487	507	526	546	565	584	604	623

One tail 1% Two tail 2%

n \ m	1	2	3	4	5	6	7	8	9	10	11	12	13	14	15	16	17	18	19	20	21	22	23	24	25	26	27	28	29	30
1																														
2									18	20	22	24	26	28	30	31	33	35	37	39	41	43	44	46	48	50	52	54	56	58
3					15	18	21	23	26	29	31	34	36	39	42	44	47	50	52	55	57	60	63	65	68	70	73	76	78	81
4				16	20	23	26	30	33	36	40	43	46	50	53	56	60	63	66	70	73	76	79	83	86	89	93	96	99	103
5			15	20	24	28	32	36	40	44	48	52	56	60	64	68	72	76	80	84	88	92	96	100	104	108	112	116	120	124
6			18	23	28	33	37	42	47	51	56	61	65	70	75	79	84	89	93	98	102	107	112	116	121	126	130	135	139	144
7			21	26	32	37	43	48	53	59	64	69	75	80	85	91	96	101	106	112	117	122	128	133	138	143	149	154	159	164
8			23	30	36	42	48	54	60	66	72	78	84	90	96	102	108	114	120	125	131	137	143	149	155	161	167	173	178	184
9		18	26	33	40	47	53	60	67	73	80	87	93	100	106	113	119	126	133	139	146	152	159	165	172	178	185	191	198	204
10		20	29	36	44	51	59	66	73	81	88	95	102	110	117	124	131	138	145	153	160	167	174	181	188	195	203	210	217	224
11		22	31	40	48	56	64	72	80	88	96	104	112	119	127	135	143	151	158	166	174	182	189	197	205	213	220	228	236	244
12		24	34	43	52	61	69	78	87	95	104	112	121	129	138	146	154	163	171	180	188	196	205	213	221	230	238	246	255	263
13		26	36	46	56	65	75	84	93	102	112	121	130	139	148	157	166	175	184	193	202	211	220	229	238	247	256	265	274	283
14		28	39	50	60	70	80	90	100	110	119	129	139	148	158	168	177	187	197	206	216	225	235	245	254	264	273	283	292	302
15		30	42	53	64	75	85	96	106	117	127	138	148	158	168	179	189	199	209	220	230	240	250	260	270	281	291	301	311	321
16		31	44	56	68	79	91	102	113	124	135	146	157	168	179	190	200	211	222	233	244	254	265	276	287	297	308	319	330	340
17		33	47	60	72	84	96	108	119	131	143	154	166	177	189	200	212	223	235	246	257	269	280	292	303	314	326	337	348	360
18		35	50	63	76	89	101	114	126	138	151	163	175	187	199	211	223	235	247	259	271	283	295	307	319	331	343	355	367	379
19		37	52	66	80	93	106	120	133	145	158	171	184	197	209	222	235	247	260	273	285	298	310	323	335	348	360	373	385	398
20		39	55	70	84	98	112	125	139	153	166	180	193	206	220	233	246	259	273	286	299	312	325	338	351	365	378	391	404	417
21		41	57	73	88	102	117	131	146	160	174	188	202	216	230	244	257	271	285	299	313	326	340	354	368	381	395	409	422	436
22		43	60	76	92	107	122	137	152	167	182	196	211	225	240	254	269	283	298	312	326	341	355	369	384	398	412	427	441	455
23		44	63	79	96	112	128	143	159	174	189	205	220	235	250	265	280	295	310	325	340	355	370	385	400	415	430	444	459	474
24		46	65	83	100	116	133	149	165	181	197	213	229	245	260	276	292	307	323	338	354	369	385	400	416	431	447	462	478	493
25		48	68	86	104	121	138	155	172	188	205	221	238	254	270	287	303	319	335	351	368	384	400	416	432	448	464	480	496	512
26		50	70	89	108	126	143	161	178	195	213	230	247	264	281	297	314	331	348	365	381	398	415	431	448	465	481	498	515	531
27		52	73	93	112	130	149	167	185	203	220	238	256	273	291	308	326	343	360	378	395	412	430	447	464	481	499	516	533	550
28		54	76	96	116	135	154	173	191	210	228	246	265	283	301	319	337	355	373	391	409	427	444	462	480	498	516	534	551	569
29		56	78	99	120	139	159	178	198	217	236	255	274	292	311	330	348	367	385	404	422	441	459	478	496	515	533	551	570	588
30		58	81	103	124	144	164	184	204	224	244	263	283	302	321	340	360	379	398	417	436	455	474	493	512	531	550	569	588	607

CRITICAL VALUES OF THE MANN-WHITNEY STATISTIC

Since U is discrete, exact significance levels cannot in general be achieved. The critical values given are those whose significance levels are nearest to those stated.

One tail 2.5% Two tail 5%

n \ m	1	2	3	4	5	6	7	8	9	10	11	12	13	14	15	16	17	18	19	20	21	22	23	24	25	26	27	28	29	30
1																				20	21	22	23	24	25	26	27	28	29	30
2					10	12	14	16	18	19	21	23	25	26	28	30	32	34	35	37	39	41	42	44	46	48	50	51	53	55
3				12	15	17	19	22	24	27	29	32	34	37	39	42	44	47	49	52	54	57	59	62	64	67	69	72	74	76
4			12	15	18	22	25	28	31	34	37	40	44	47	50	53	56	59	62	66	69	72	75	78	81	84	87	91	94	97
5		10	15	18	22	26	30	34	38	41	45	49	53	56	60	64	68	72	75	79	83	87	90	94	98	102	105	109	113	117
6		12	17	22	26	31	35	40	44	48	53	57	62	66	70	75	79	84	88	92	97	101	106	110	114	119	123	128	132	136
7		14	19	25	30	35	40	45	50	55	60	66	71	76	81	86	91	96	101	106	111	116	121	126	131	136	141	146	151	156
8		16	22	28	34	40	45	51	57	62	68	74	79	85	91	96	102	108	113	119	124	130	136	141	147	152	158	164	169	175
9		18	24	31	38	44	50	57	63	69	76	82	88	94	101	107	113	119	126	132	138	144	151	157	163	169	175	182	188	194
10		19	27	34	41	48	55	62	69	76	83	90	97	104	111	117	124	131	138	145	152	158	165	172	179	186	193	199	206	213
11		21	29	37	45	53	60	68	76	83	91	98	106	113	121	128	135	143	150	158	165	173	180	187	195	202	210	217	224	232
12		23	32	40	49	57	66	74	82	90	98	106	114	122	130	139	147	155	163	171	179	187	195	203	211	219	227	235	243	251
13		25	34	44	53	62	71	79	88	97	106	114	123	132	140	149	158	166	175	183	192	201	209	218	226	235	244	252	261	269
14		26	37	47	56	66	76	85	94	104	113	122	132	141	150	159	169	178	187	196	205	215	224	233	242	251	261	270	279	288
15		28	39	50	60	70	81	91	101	111	121	130	140	150	160	170	180	189	199	209	219	229	238	248	258	268	277	287	297	307
16		30	42	53	64	75	86	96	107	117	128	139	149	159	170	180	191	201	211	222	232	243	253	263	274	284	294	305	315	325
17		32	44	56	68	79	91	102	113	124	135	147	158	169	180	191	202	213	224	235	246	256	267	278	289	300	311	322	333	344
18		34	47	59	72	84	96	108	119	131	143	155	166	178	189	201	213	224	236	247	259	270	282	293	305	316	328	339	351	362
19		35	49	62	75	88	101	113	126	138	150	163	175	187	199	211	224	236	248	260	272	284	296	308	320	333	345	357	369	381
20	20	37	52	66	79	92	106	119	132	145	158	171	183	196	209	222	235	247	260	273	285	298	311	323	336	349	361	374	387	399
21	21	39	54	69	83	97	111	124	138	152	165	179	192	205	219	232	246	259	272	285	299	312	325	338	352	365	378	391	404	418
22	22	41	57	72	87	101	116	130	144	158	173	187	201	215	229	243	256	270	284	298	312	326	340	353	367	381	395	409	422	436
23	23	42	59	75	90	106	121	136	151	165	180	195	209	224	238	253	267	282	296	311	325	340	354	368	383	397	411	426	440	454
24	24	44	62	78	94	110	126	141	157	172	187	203	218	233	248	263	278	293	308	323	338	353	368	383	398	413	428	443	458	473
25	25	46	64	81	98	114	131	147	163	179	195	211	226	242	258	274	289	305	320	336	352	367	383	398	414	429	445	460	476	491
26	26	48	67	84	102	119	136	152	169	186	202	219	235	251	268	284	300	316	333	349	365	381	397	413	429	445	461	477	493	510
27	27	50	69	87	105	123	141	158	175	193	210	227	244	261	277	294	311	328	345	361	378	395	411	428	445	461	478	495	511	528
28	28	51	72	91	109	128	146	164	182	199	217	235	252	270	287	305	322	339	357	374	391	409	426	443	460	477	495	512	529	546
29	29	53	74	94	113	132	151	169	188	206	224	243	261	279	297	315	333	351	369	387	404	422	440	458	476	493	511	529	547	564
30	30	55	76	97	117	136	156	175	194	213	232	251	269	288	307	325	344	362	381	399	418	436	454	473	491	510	528	546	564	583

One tail 5% Two tail 10%

n \ m	1	2	3	4	5	6	7	8	9	10	11	12	13	14	15	16	17	18	19	20	21	22	23	24	25	26	27	28	29	30
1										10	11	12	13	14	15	16	17	18	19	20	21	22	23	24	25	26	27	28	28	29
2			6	8	10	12	13	15	17	18	20	22	23	25	27	28	30	32	33	35	37	39	40	42	44	45	47	49	50	52
3		6	9	11	14	16	18	21	23	25	28	30	32	35	37	39	42	44	46	49	51	53	56	58	60	63	65	67	70	72
4		8	11	14	17	20	23	26	29	32	35	38	41	44	47	50	53	56	59	62	65	68	71	74	77	80	83	85	88	91
5		10	14	17	21	25	28	32	35	39	43	46	50	53	57	61	64	68	71	75	78	82	86	89	93	96	100	103	107	111
6		12	16	20	25	29	33	37	42	46	50	54	58	63	67	71	75	79	83	88	92	96	100	104	109	113	117	121	125	129
7		13	18	23	28	33	38	43	48	52	57	62	67	72	76	81	86	91	96	100	105	110	115	119	124	129	134	138	143	148
8		15	21	26	32	37	43	48	54	59	64	70	75	81	86	91	97	102	108	113	118	124	129	134	140	145	150	156	161	167
9		17	23	29	35	42	48	54	60	66	72	78	84	90	96	102	108	114	120	125	131	137	143	149	155	161	167	173	179	185
10	10	18	25	32	39	46	52	59	66	72	79	86	92	99	105	112	118	125	131	138	145	151	158	164	171	177	184	190	197	203
11	11	20	28	35	43	50	57	64	72	79	86	93	100	108	115	122	129	136	143	150	158	165	172	179	186	193	200	207	214	221
12	12	22	30	38	46	54	62	70	78	86	93	101	109	117	124	132	140	147	155	163	171	178	186	194	201	209	217	224	232	240
13	13	23	32	41	50	58	67	75	84	92	100	109	117	125	134	142	150	159	167	175	183	192	200	208	217	225	233	241	250	258
14	14	25	35	44	53	63	72	81	90	99	108	117	125	134	143	152	161	170	179	188	196	205	214	223	232	241	249	258	267	276
15	15	27	37	47	57	67	76	86	96	105	115	124	134	143	153	162	172	181	190	200	209	219	228	238	247	256	266	275	285	294
16	16	28	39	50	61	71	81	91	102	112	122	132	142	152	162	172	182	192	202	212	222	232	242	252	262	272	282	292	302	312
17	17	30	42	53	64	75	86	97	108	118	129	140	150	161	172	182	193	203	214	225	235	246	256	267	277	288	298	309	319	330
18	18	32	44	56	68	79	91	102	114	125	136	147	159	170	181	192	203	215	226	237	248	259	270	281	292	303	315	326	337	348
19	19	33	46	59	71	83	96	108	120	131	143	155	167	179	190	202	214	226	237	249	261	272	284	296	307	319	331	342	354	366
20	20	35	49	62	75	88	100	113	125	138	150	163	175	188	200	212	225	237	249	261	274	286	298	310	323	335	347	359	371	384
21	21	37	51	65	78	92	105	118	131	145	158	171	183	196	209	222	235	248	261	274	286	299	312	325	338	350	363	376	389	402
22	22	39	53	68	82	96	110	124	137	151	165	178	192	205	219	232	246	259	272	286	299	313	326	339	353	366	379	393	406	419
23	23	40	56	71	86	100	115	129	143	158	172	186	200	214	228	242	256	270	284	298	312	326	340	354	368	382	396	409	423	437
24	24	42	58	74	89	104	119	134	149	164	179	194	208	223	238	252	267	281	296	310	325	339	354	368	383	397	412	426	441	455
25	25	44	60	77	93	109	124	140	155	171	186	201	217	232	247	262	277	292	307	323	338	353	368	383	398	413	428	443	458	473
26	26	45	63	80	96	113	129	145	161	177	193	209	225	241	256	272	288	303	319	335	350	366	382	397	413	428	444	460	475	491
27	27	47	65	83	100	117	134	150	167	184	200	217	233	249	266	282	298	315	331	347	363	379	396	412	428	444	460	476	492	509
28	28	49	67	85	103	121	138	156	173	190	207	224	241	258	275	292	309	326	342	359	376	393	409	426	443	460	476	493	510	526
29	28	50	70	88	107	125	143	161	179	197	214	232	250	267	285	302	319	337	354	371	389	406	423	441	458	475	492	510	527	544
30	29	52	72	91	111	129	148	167	185	203	221	240	258	276	294	312	330	348	366	384	402	419	437	455	473	491	509	526	544	562

TABLE 15 RANDOM DIGITS

The table gives 2500 random digits, from 0 to 9, arranged for convenience in blocks of 5.

87024	74221	69721	44518	58804	04860	18127	16855	61558	15430
04852	03436	72753	99836	37513	91341	53517	92094	54386	44563
33592	45845	52015	72030	23071	92933	84219	39455	57792	14216
68121	53688	56812	34869	28573	51079	94677	23993	88241	97735
25062	10428	43930	69033	73395	83469	25990	12971	73728	03856
78183	44396	11064	92153	96293	00825	21079	78337	19739	13684
70209	23316	32828	00927	61841	64754	91125	01206	06691	50868
94342	91040	94035	02650	36284	91162	07950	36178	42536	49869
92503	29854	24116	61149	49266	82303	54924	58251	23928	20703
71646	57503	82416	22657	72359	30085	13037	39608	77439	49318
51809	70780	41544	27828	84321	07714	25865	97896	01924	62028
88504	21620	07292	71021	80929	45042	08703	45894	24521	49942
33186	49273	87542	41086	29615	81101	43707	87031	36101	15137
40068	35043	05280	62921	30122	65119	40512	26855	40842	83244
76401	68461	20711	12007	19209	28259	49820	76415	51534	63574
47014	93729	74235	47808	52473	03145	92563	05837	70023	33169
67147	48017	90741	53647	55007	36607	29360	83163	79024	26155
86987	62924	93157	70947	07336	49541	81386	26968	38311	99885
58973	47026	78574	08804	22960	32850	67944	92303	61216	72948
71635	86749	40369	94639	40731	54012	03972	98581	45604	34885
60971	54212	32596	03052	84150	36798	62635	26210	95685	87089
06599	60910	66315	96690	19039	39878	44688	65146	02482	73130
89960	27162	66264	71024	18708	77974	40473	87155	35834	03114
03930	56898	61900	44036	90012	17673	54167	82396	39468	49566
31338	28729	02095	07429	35718	86882	37513	51560	08872	33717
29782	33287	27400	42915	49914	68221	56088	06112	95481	30094
68493	88796	94771	89418	62045	40681	15941	05962	44378	64349
42534	31925	94158	90197	62874	53659	33433	48610	14698	54761
76126	41049	43363	52461	00552	93352	58497	16347	87145	73668
80434	73037	69008	36801	25520	14161	32300	04187	80668	07499
81301	39731	53857	19690	39998	49829	12399	70867	44498	17385
54521	42350	82908	51212	70208	39891	64871	67448	42988	32600
82530	22869	87276	06678	36873	61198	87748	07531	29592	39612
81338	64309	45798	42954	95565	02789	83017	82936	67117	17709
58264	60374	32610	17879	96900	68029	06993	84288	35401	56317
77023	46829	21332	77383	15547	29332	77698	89878	20489	71800
29750	59902	78110	59018	87548	10225	15774	70778	56086	08117
08288	38411	69886	64918	29055	87607	37452	38174	31431	46173
93908	94810	22057	94240	89918	16561	92716	66461	22337	64718
06341	25883	42574	80202	57287	95120	69332	19036	43326	98697
23240	94741	55622	79479	34606	51079	09476	10695	49618	63037
96370	19171	40441	05002	33165	28693	45027	73791	23047	32976
97050	16194	61095	26533	81738	77032	60551	31605	95212	81078
40833	12169	10712	78345	48236	45086	61654	94929	69169	70561
95676	13582	25664	60838	88071	50052	63188	50346	65618	17517
28030	14185	13226	99566	45483	10079	22945	23903	11695	10694
60202	32586	87466	83357	95516	31258	66309	40615	30572	60842
46530	48755	02308	79508	53422	50805	08896	06963	93922	99423
53151	95839	01745	46462	81463	28669	60179	17880	75875	34562
80272	64398	88249	06792	98424	66842	49129	98939	34173	49883

TABLE 16 NEGATIVE EXPONENTIAL FUNCTION

The table gives the values of the function $f(x) = e^{-x}$.

SUBTRACT

x	.00	.01	.02	.03	.04	.05	.06	.07	.08	.09	1	2	3	4	5	6	7	8	9
0.0	1.0000	.99005	.98020	.97045	.96079	.95123	.94176	.93239	.92312	.91393	95	190	286	381	476	571	666	760	855
0.1	.90484	.89583	.88692	.87810	.86936	.86071	.85214	.84366	.83527	.82696	86	172	258	344	431	516	602	688	774
0.2	.81873	.81058	.80252	.79453	.78663	.77880	.77105	.76338	.75578	.74826	78	156	234	312	389	467	545	623	700
0.3	.74082	.73345	.72615	.71892	.71177	.70469	.69768	.69073	.68386	.67706	70	141	211	282	352	423	493	563	633
0.4	.67032	.66365	.65705	.65051	.64404	.63763	.63128	.62500	.61878	.61263	63	127	191	255	319	382	446	510	573
0.5	.60653	.60050	.59452	.58860	.58275	.57695	.57121	.56553	.55990	.55433	57	115	173	231	288	346	404	461	518
0.6	.54881	.54335	.53794	.53259	.52729	.52205	.51685	.51171	.50662	.50158	52	104	157	209	261	313	365	417	469
0.7	.49659	.49164	.48675	.48191	.47711	.47237	.46767	.46301	.45841	.45384	47	94	142	189	236	283	330	377	424
0.8	.44933	.44486	.44043	.43605	.43171	.42741	.42316	.41895	.41478	.41066	42	85	128	171	214	256	299	341	384
0.9	.40657	.40252	.39852	.39455	.39063	.38674	.38289	.37908	.37531	.37158	38	77	116	155	193	232	270	309	347
1.0	.36788	.36422	.36059	.35701	.35345	.34994	.34646	.34301	.33960	.33622	35	70	105	140	175	210	245	279	314
1.1	.33287	.32956	.32628	.32303	.31982	.31664	.31349	.31037	.30728	.30422	31	63	95	126	158	190	221	253	284
1.2	.30119	.29820	.29523	.29229	.28938	.28650	.28365	.28083	.27804	.27527	28	57	86	114	143	172	200	229	257
1.3	.27253	.26982	.26714	.26448	.26185	.25924	.25666	.25411	.25158	.24908	26	52	77	103	129	155	181	207	233
1.4	.24660	.24414	.24171	.23931	.23693	.23457	.23224	.22993	.22764	.22537	23	47	70	94	117	140	164	187	211
1.5	.22313	.22091	.21871	.21654	.21438	.21225	.21014	.20805	.20598	.20393	21	42	63	85	106	127	148	169	190
1.6	.20190	.19989	.19790	.19593	.19398	.19205	.19014	.18825	.18637	.18452	19	38	57	76	96	115	134	153	172
1.7	.18268	.18087	.17907	.17728	.17552	.17377	.17204	.17033	.16864	.16696	17	34	52	69	87	104	121	139	156
1.8	.16530	.16365	.16203	.16041	.15882	.15724	.15567	.15412	.15259	.15107	15	31	47	63	78	94	110	125	141
1.9	.14957	.14808	.14661	.14515	.14370	.14227	.14086	.13946	.13807	.13670	14	28	42	57	71	85	99	113	127
2.0	.13534	.13399	.13266	.13134	.13003	.12873	.12745	.12619	.12493	.12369	12	25	38	51	64	77	90	102	115
2.1	.12246	.12124	.12003	.11884	.11765	.11648	.11533	.11418	.11304	.11192	11	23	35	46	58	69	81	93	104
2.2	.11080	.10970	.10861	.10753	.10646	.10540	.10435	.10331	.10228	.10127	10	21	31	42	52	63	73	84	94
2.3	.10026	.09926	.09827	.09730	.09633	.09537	.09442	.09348	.09255	.09163	9	19	28	38	47	57	66	76	85
2.4	.09072	.08982	.08892	.08804	.08716	.08629	.08543	.08458	.08374	.08291	8	17	25	34	43	51	60	69	77
2.5	.08208	.08127	.08046	.07966	.07887	.07808	.07730	.07654	.07577	.07502	7	15	23	31	39	46	54	62	70
2.6	.07427	.07353	.07280	.07208	.07136	.07065	.06995	.06925	.06856	.06788	7	14	21	28	35	42	49	56	63
2.7	.06721	.06654	.06587	.06522	.06457	.06393	.06329	.06266	.06204	.06142	6	12	19	25	32	38	44	51	57
2.8	.06081	.06020	.05961	.05901	.05843	.05784	.05727	.05670	.05613	.05558	5	11	17	23	28	34	40	46	52
2.9	.05502	.05448	.05393	.05340	.05287	.05234	.05182	.05130	.05079	.05029	5	10	15	20	26	31	36	41	47
3.0	.04979	.04929	.04880	.04832	.04783	.04736	.04689	.04642	.04596	.04550	4	9	14	18	23	28	33	37	42
3.1	.04505	.04460	.04416	.04372	.04328	.04285	.04243	.04200	.04159	.04117	4	8	12	17	21	25	30	34	38
3.2	.04076	.04036	.03996	.03956	.03916	.03877	.03839	.03801	.03763	.03725	3	7	11	15	19	23	27	31	34
3.3	.03688	.03652	.03615	.03579	.03544	.03508	.03474	.03439	.03405	.03371	3	7	10	14	17	21	24	28	31
3.4	.03337	.03304	.03271	.03239	.03206	.03175	.03143	.03112	.03081	.03050	3	6	9	12	15	19	22	25	28
3.5	.03020	.02990	.02960	.02930	.02901	.02872	.02844	.02816	.02788	.02760	2	5	8	11	14	17	20	22	25
3.6	.02732	.02705	.02678	.02652	.02625	.02599	.02573	.02548	.02522	.02497	2	5	7	10	13	15	18	20	23
3.7	.02472	.02448	.02423	.02399	.02375	.02352	.02328	.02305	.02282	.02260	2	4	7	9	11	14	16	18	21
3.8	.02237	.02215	.02193	.02171	.02149	.02128	.02107	.02086	.02065	.02045	2	4	6	8	10	12	14	17	19
3.9	.02024	.02004	.01984	.01964	.01945	.01925	.01906	.01887	.01869	.01850	1	3	5	7	9	11	13	15	17
4.0	.01832	.01813	.01795	.01777	.01760	.01742	.01725	.01708	.01691	.01674	1	3	5	6	8	10	12	13	15
4.1	.01657	.01641	.01624	.01608	.01592	.01576	.01561	.01545	.01530	.01515	1	3	4	6	7	9	11	12	14
4.2	.01500	.01485	.01470	.01455	.01441	.01426	.01412	.01398	.01384	.01370	1	2	4	5	7	8	9	11	12
4.3	.01357	.01343	.01330	.01317	.01304	.01291	.01278	.01265	.01253	.01240	1	2	3	5	6	7	9	10	11
4.4	.01228	.01216	.01203	.01191	.01180	.01168	.01156	.01145	.01133	.01122	1	2	3	4	5	7	8	9	10
4.5	.01111	.01100	.01089	.01078	.01067	.01057	.01046	.01036	.01025	.01015	1	2	3	4	5	6	7	8	9
4.6	.01005	.00995	.00985	.00975	.00966	.00956	.00947	.00937	.00928	.00919	0	1	2	3	4	5	6	7	8
4.7	.00910	.00900	.00892	.00883	.00874	.00865	.00857	.00848	.00840	.00831	0	1	2	3	4	5	6	6	7
4.8	.00823	.00815	.00807	.00799	.00791	.00783	.00775	.00767	.00760	.00752	0	1	2	3	3	4	5	6	7
4.9	.00745	.00737	.00730	.00723	.00715	.00708	.00701	.00694	.00687	.00681	0	1	2	2	3	4	4	5	6

OTHER PUBLICATIONS FROM RND

The following Advanced level textbooks on Statistics are published by RND:

Introductory Statistics ISBN 0 - 9506719 - 1 - 6
Volume 1 - Probability and Distribution Theory

Worked Examples in Probability and Distribution Theory ISBN 0 - 9506719 - 2 - 4

Introductory Statistics ISBN 0 - 9506719 - 3 - 2
Volume 2 - Statistical Inference

Worked Examples in Statistical Inference ISBN 0 - 9506719 - 4 - 0

Booklets of model answers to examination papers are available for the following boards and levels :

Welsh Joint Education Committee

Advanced Level

- Computer Science
- Mathematics
- Physics

G.C.S.E.

- Mathematics
- Physics

Associated Examining Board

Advanced Level

- Accountancy
- Applied Mathematics
- Computer Science
- Economics
- Pure Mathematics
- Statistics

A set of Revision Notes for G.C.S.E. Physics is also available.

Further details can be obtained from the Publishers.